Shortcut Algebra II

Shortcut Algebra II

Andrew Marx

PUBLISHING

New York

This publication is designed to provide accurate and authoritative information in regard to the subject matter covered. It is sold with the understanding that the publisher is not engaged in rendering legal, accounting, or other professional service. If legal advice or other expert assistance is required, the services of a competent professional should be sought.

Editorial Director: Jennifer Farthing
Editor: Cynthia Ierardo
Production Editor: Dominique Polfliet
Production Artist: John Christensen
Cover Designer: Carly Schnur

Published by Kaplan Publishing, a division of Kaplan, Inc.
888 Seventh Ave.
New York, NY 10106

Printed in the United States of America

June 2007

09 10 11   10 9 8 7 6 5 4

ISBN 13: 978-1-4195-9315-4
ISBN 10: 1-4195-9315-3

TABLE OF CONTENTS

INTRODUCTION TO SHORTCUT ALGEBRA II

If you have picked up this book, you might be one of the many people who think algebra is an intimidating topic, but who needs more advanced knowledge of the subject. Maybe you're an algebra student who has a handle on the basics but needs some additional guidance in a high school or college class. Maybe you're a student gearing up for a standardized test in math. Maybe you're a professional who needs to apply more advanced mathematics in your work.

If you are one of these people, this book is designed just for you! *Shortcut Algebra II* offers an easy-to-understand approach that will guide you through the maze of problems and proofs that comprise basic algebra. With over 200 step-by-step examples and practice questions, you'll be well on your way to feeling confident and at-ease with this challenging subject.

Algebra II involves a small number of concepts and skills tested in different ways. Once you realize that, you'll appreciate the importance of setting up problems so that solving them becomes straightforward. Once a problem is set up properly, it's just a matter of going through standard steps to get to the solution.

Shortcut Algebra II begins with a review of common algebra topics that are the foundation for working with functions, which is mainly what intermediate-level algebra is all about.

To use the book to the fullest advantage, start by taking the Diagnostic Quiz. Following the quiz, the Diagnostic Correlation Chart and detailed answer explanations will help you identify your weak areas and will direct you to the appropriate chapters in the book for review. Depending on your needs, you may choose to skip directly to those chapters or to work through the whole book from beginning to end.

If you have enough time to do so, we recommend that you work through the entire book, because many chapters build on information presented in previous chapters. It would be to your advantage to review all the concepts and practice all the problems. For example, to understand quadratic and polynomial functions, you need to understand the basics of functions.

To understand probability, you need to understand combinations and permutations.

Each core chapter is structured to identify key concepts and outline steps that will help you solve the most common types of questions. You'll learn how to apply those steps to real problems. As you wend your way to a solution, detailed explanations will walk you through problem-solving techniques and useful strategies. Each chapter concludes with a ten-problem quiz, which will help you to evaluate yourself and apply your understanding. In-depth explanatory solutions are provided for the quiz problems as well, so that you can check your work and target any areas where you may still need some review.

By the time you reach the end of *Shortcut Algebra II*, we're confident that you will see algebra in a whole new light—and you'll be amazed that it took such a short time to get from where you started to a clear understanding of the basics. You will be well on your way to mastering the essential skills and concepts that you need to succeed.

Good luck—and enjoy the shortcut!

Diagnostic Test

This brief test covers every major topic in *Shortcut Algebra II*. It will help you identify your areas of strength and the areas that will require extra attention and review.

After you have scored your test, you can use the chart included at the end to analyze your results. The chart matches every question with a section of the book. You can use it to jump ahead if you need a quick review.

Keep in mind that the book covers much more than what can be tested here. This diagnostic test can still give you a good idea of how much time you'll need to spend with each chapter.

1. If $f(x) = x^2 + 6$, then $f(-4) =$
    - (A) −10
    - (B) −2
    - (C) 4
    - (D) 10
    - (E) 22

2. Which function has a range of real numbers greater than or equal to −2?
    - (A) $f(x) = -3x^2 + 2$
    - (B) $f(x) = -2 - 2x^2$
    - (C) $f(x) = 2 - 2x^2$
    - (D) $f(x) = 2x^2 - 4$
    - (E) $f(x) = 3x^2 - 2$

3. What is the zero of $f(a) = 6a - 11$?
    - (A) $-\dfrac{6}{11}$
    - (B) $-\dfrac{1}{6}$
    - (C) $\dfrac{1}{11}$
    - (D) $\dfrac{6}{11}$
    - (E) $\dfrac{11}{6}$

4. What is the y-intercept of the graph of $f(x) = \dfrac{3x - 5}{7}$?

   (A) -5

   (B) $-\dfrac{5}{7}$

   (C) $\dfrac{5}{3}$

   (D) 3

   (E) 7

5. The graph below is a straight line.

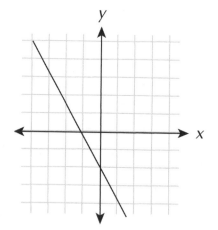

The graph represents which function?

   (A) $f(x) = -\dfrac{7x}{4} - 2$

   (B) $f(x) = \dfrac{7x}{4} - 1$

   (C) $f(x) = \dfrac{4x}{7} - 2$

   (D) $f(x) = 2x - \dfrac{7}{4}$

   (E) $f(x) = -2x + \dfrac{4}{7}$

6. What are the zeroes of $f(x) = x^2 - 11x - 42$?

   (A) -9 and 2

   (B) -6 and 5

   (C) -4 and 7

   (D) -3 and 14

   (E) -2 and 21

7. Which quadratic equation has two imaginary roots?

   (A) $x^2 - 4x - 5 = 0$

   (B) $x^2 - 3x + 5 = 0$

   (C) $x^2 + 6x - 8 = 0$

   (D) $2x^2 + 8x - 10 = 0$

   (E) $3x^2 - x - 1 = 0$

8. What is the range of $f(x) = x^2 - 8x + 19$?

   (A) All real numbers less than or equal to –4
   (B) All real numbers greater than or equal to 3
   (C) All real numbers less than or equal to 3
   (D) All real numbers greater than or equal to 4
   (E) All real numbers less than or equal to 4

9. What are the zeroes of $f(x) = x^3 + 3x^2 - 6x - 8$?

   (A) –4, –1, and 2
   (B) –4, 1, and –2
   (C) –4, 1, and 2
   (D) 4, –1, and –2
   (E) 4, –1, and 2

10. What is the domain of
    $f(x) = \dfrac{x^3 + 8}{x^2 - 9}$?

    (A) All real numbers other than –3 and 3
    (B) All real numbers other than –2 and 3
    (C) All real numbers less than –3 or greater than 3
    (D) All real numbers greater than –2
    (E) All real numbers other than –2

11. $\begin{bmatrix} 3 & -4 & 5 \\ 5 & 2 & -7 \\ 8 & -8 & 1 \end{bmatrix} + \begin{bmatrix} 0 & 6 & 5 \\ 1 & -3 & 3 \\ -9 & -7 & -1 \end{bmatrix} =$

    (A) $\begin{bmatrix} 0 & -2 & 5 \\ 6 & -1 & -4 \\ -1 & -15 & 1 \end{bmatrix}$

    (B) $\begin{bmatrix} 3 & -2 & 5 \\ 6 & -1 & -10 \\ -1 & -1 & -1 \end{bmatrix}$

    (C) $\begin{bmatrix} 3 & 2 & 10 \\ 6 & -1 & -4 \\ -1 & -15 & 0 \end{bmatrix}$

    (D) $\begin{bmatrix} 3 & -2 & 10 \\ 5 & -1 & -4 \\ 1 & -1 & -1 \end{bmatrix}$

    (E) $\begin{bmatrix} 3 & -2 & 10 \\ 6 & -5 & -10 \\ 1 & -15 & 0 \end{bmatrix}$

12. $\begin{bmatrix} 2 & 3 \end{bmatrix} \times \begin{bmatrix} 3 & -2 & 6 \\ 5 & -1 & -3 \end{bmatrix} =$

    (A) $\begin{bmatrix} 21 & -7 & 3 \end{bmatrix}$

    (B) $\begin{bmatrix} 21 \\ -7 \\ 3 \end{bmatrix}$

    (C) $\begin{bmatrix} 14 \\ -3 \end{bmatrix}$

    (D) $\begin{bmatrix} 6 & 15 \\ -4 & -3 \\ 12 & -9 \end{bmatrix}$

    (E) $\begin{bmatrix} 6 & -4 & 12 \\ 15 & -3 & -9 \end{bmatrix}$

13. What is the equation of the following circle?

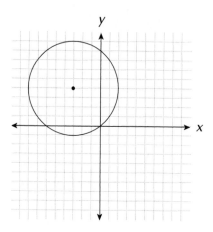

(A) $(x-3)^2 + (y+4)^2 = 5$

(B) $(x+3)^2 + (y-4)^2 = 5$

(C) $(x-3)^2 + (y+4)^2 = 10$

(D) $(x-3)^2 + (y+4)^2 = 25$

(E) $(x+3)^2 + (y-4)^2 = 25$

14. What is the diameter of the circle $(x-8)^2 + (y+7)^2 = 49$?

(A) 7

(B) 8

(C) 14

(D) 16

(E) 49

15. What is the length of the major or horizontal axis of the ellipse $\dfrac{x^2}{81} + \dfrac{y^2}{9} = 1$?

(A) 3

(B) 8

(C) 9

(D) 16

(E) 18

16. What is the approximate value of $3.5^{1.5}$?

(A) $1.87\ (3.5^{1.5})$

(B) 2.52 raised log of 3.5 to 1.5 power

(C) 5.25

(D) 6.54

(E) 12.25

17. If $x^{1.5} = 343$, then $x =$

(A) 7

(B) 14

(C) 21

(D) 28

(E) 49

18. If $256^a = 1,048,576$, then $a =$

(A) 1.5

(B) 2.5

(C) 3.25

(D) 3.5

(E) 3.75

19. The right triangle below includes angle *A*.

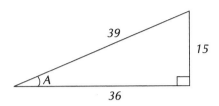

What is the cosine of *A*?

(A) $\dfrac{5}{13}$

(B) $\dfrac{5}{12}$

(C) $\dfrac{12}{13}$

(D) $\dfrac{12}{5}$

(E) $\dfrac{13}{5}$

20. The right triangle below has labeled side measures.

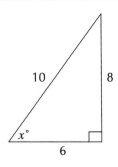

What is the value of *x* to the nearest tenth?

(A) 36.9

(B) 38.6

(C) 41.4

(D) 53.1

(E) 57.8

21. A triangle on the coordinate plane has a vertex with coordinates (–4, 8). Which of the following are the coordinates of a vertex of the triangle after it is rotated 180° around the origin?

(A) (–8, 4)

(B) (4, –8)

(C) (4, 8)

(D) (8, –4)

(E) (8, 4)

22. The coordinate plane below includes a triangle.

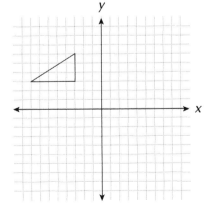

Which of the following is the translation of the previous figure by (6, −2)?

(A)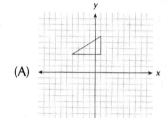

(B)

(C)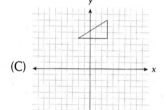

(D)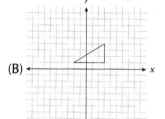

(E)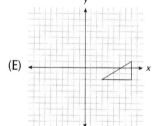

23. The figures on the coordinate plane below are reflections over what line?

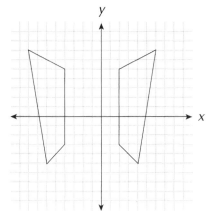

(A) *x*-axis

(B) *x* = −1

(C) *x* = −2

(D) *y*-axis

(E) *y* = 1

24. What is the next number in the sequence 8, 24, 72, __

(A) 96

(B) 120

(C) 144

(D) 184

(E) 216

25. If the third and fourth terms in an arithmetic series are 34 and 41, what is the first term?

(A) 13

(B) 20

(C) 27

(D) 48

(E) 55

26. What is the sum of the geometric sequence

$36, 12, 4, \frac{4}{3}, \dots ?$

(A) 54

(B) 56

(C) 60

(D) 68

(E) 72

27. The digits 2, 3, 4, and 5 can be arranged to form how many four-digit numbers?

(A) 6

(B) 12

(C) 24

(D) 60

(E) 120

28. From a group of five employees, three will be selected at random to work on a special committee. How many possible selections of three employees are there?

    (A) 10

    (B) 20

    (C) 40

    (D) 60

    (E) 120

29. What is the probability of three coin tosses all landing on heads?

    (A) $\dfrac{1}{16}$

    (B) $\dfrac{1}{12}$

    (C) $\dfrac{1}{8}$

    (D) $\dfrac{1}{6}$

    (E) $\dfrac{1}{4}$

30. What is the probability of rolling a total of 10 with a pair of six-sided dice?

    (A) $\dfrac{1}{12}$

    (B) $\dfrac{1}{10}$

    (C) $\dfrac{1}{9}$

    (D) $\dfrac{1}{6}$

    (E) $\dfrac{1}{4}$

ANSWER KEY

1. E	9. A	17. E	25. B
2. E	10. A	18. B	26. A
3. E	11. C	19. C	27. C
4. B	12. A	20. D	28. A
5. A	13. E	21. B	29. C
6. D	14. C	22. B	30. A
7. B	15. C	23. B	
8. B	16. D	24. E	

ANSWER EXPLANATIONS

1. E

This question deals with an important use of functions, which are introduced in Chapter 1. Here, your function $f(x)$ is said to be $x^2 + 6$. Finding the value of $f(-4)$ is just a matter of evaluating that expression when $x = -4$. If $x = -4$, $x^2 + 6 = (-4)^2 + 6 = 16 + 6 = 22$. Choice (C) could be the result of squaring the sum of -4 and 6. You might have gotten (A) if you got -16 instead of 16 when squaring -4.

2. E

This is another basic function question. We'll begin to talk about ranges in Chapter 1. When a question asks to find or identify a range, you need to determine the possible values $f(x)$ can have. You can look at each function, and ask whether $f(x)$ can possibly have a value less than -2. Start by looking at the variable terms: each of them includes x^2. For all values of x, $x^2 \geq 0$. Thus, for all values of x, $3x^2 \geq 0$. It follows further that $3x^2 - 2 \geq -2$. So (E) is correct.

3. E

The zero of a function, another key concept introduced in Chapter 1, is the value of the variable that gives the expression on the right side a value of 0.

So all you have to do to find the zero of $f(a) = 6a - 11$ is solve $0 = 6a - 11$ for a:

$$0 = 6a - 11$$
$$\underline{+11 \qquad +11}$$
$$\underline{11 = 6a}$$
$$6 \qquad 6$$
$$\frac{11}{6} = a$$

So (E) is correct. (D) is the result of dividing 6 by 11 instead of 11 by 6 in the last step.

4. B

This question involves a linear function. That's the subject of Chapter 2. You'll see that we treat a function such as $f(x) = \dfrac{3x - 5}{7}$ like a linear equation such as $y = \dfrac{3x - 5}{7}$. The y–intercept of a linear equation is the value of y when $x = 0$. So the y–intercept of a linear function $f(x)$ is $f(0)$. Therefore, we evaluate $\dfrac{3x - 5}{7}$ when $x = 0$:

$$\frac{3x - 5}{7} = \frac{3(0) - 5}{7} = \frac{0 - 5}{7} = -\frac{5}{7}$$

Thus, (B) is the answer. (C) might be a tempting choice because it represents the x–intercept. In terms of functions, it represents the zero of $f(x) = \dfrac{3x - 5}{7}$.

5. A

Chapter 2 also deals with identifying linear functions presented in graphs. Two pieces of information are needed to write a linear function: the slope and the y–intercept.

The slope of a line describes its direction and "steepness." It is the ratio of the difference in y–coordinates of two points on the line to the difference between their x–coordinates.

$$m = \frac{y_2 - y_1}{x_2 - x_1}$$

Two points on this graph are (–4, 5) and (0, –2). Let's plug them into the slope formula:

$$m = \frac{y_2 - y_1}{x_2 - x_1} = \frac{-2 - (5)}{0 - (-4)} = \frac{-2 - 5}{0 + 4} = -\frac{7}{4}$$

The second coordinate we used represents the *y*–intercept, the point where the line intercepts the *y*–axis. That *y*–intercept is the *y* coordinate of that point, –2.

With information, we can put the linear equation in *slope–intercept form*.

This is the form $y = mx + b$, where *m* is the slope and *b* is the *y*–intercept of the line. Here, $m = -\frac{7}{4}$ and $b = -2$.

So the function is $f(x) = -\frac{7x}{4} - 2$

6. D

This question presents another kind of function. Quadratic functions are the subject of Chapter 3. Just like in question 3, you need to solve the equation $f(x) = 0$. So you need to solve $x^2 - 11x - 42$. Unlike linear functions, a quadratic function can have two zeroes. There are two values of *x* for which $x^2 - 11x - 42$ has a value of 0. That's why each answer choice has two numbers.

$x^2 - 11x - 42$ can be factored into two binomials: $(x - 14)$ and $(x + 3)$. Since $x^2 - 11x - 42 = (x - 14) \bullet (x + 3)$, and the product of any term and 0 is 0, $x^2 - 11x - 42 = 0$ if $(x - 14) = 0$ or $(x + 3) = 0$. $(x - 14) = 0$ if $x = 14$, and $(x + 3) = 0$ if $x = -3$. So –3 and 14 are the zeroes of $f(x) = x^2 - 11x - 42$, and (D) is correct.

(E) might be a tempting choice because the numbers there also have a product of –42. They don't have a sum close to 11, though. We'll review the steps for identifying factors in Chapter 3.

7. B

Just as we talk about quadratic functions having zeroes, we can talk about the solutions of a quadratic equation, such as $x^2 - 5x + 6 = 0$ as having roots. Before Algebra II, all quadratic equations you dealt with probably had solutions you could express with real numbers. Unreal numbers are called imaginary numbers. We'll explain more in Chapter 3. For now, you can answer the question by finding what we call the discriminant. *discrepant.*

For a quadratic equation, $ax^2 + bx + c = 0$, the discriminant is $b^2 - 4ac$. If $b^2 - 4ac < 0$, then the equation has unreal or imaginary roots. To answer this question, then, we can find the discriminant of each equation:

$x^2 - 4x - 5 = 0$: $b^2 - 4ac = (-4)^2 - 4(1)(-5) = 16 - (-20) = 16 + 20 = 36$

$x^2 - 3x + 5 = 0$: $b^2 - 4ac = (-3)^2 - 4(1)(5) = 9 - 20 = -11$

$x^2 + 6x - 8 = 0$: $b^2 - 4ac = (6)^2 - 4(1)(-8) = 36 - (-32) = 36 + 32 = 68$

$2x^2 + 8x - 10 = 0$: $b^2 - 4ac = (8)^2 - 4(2)(-10) = 64 - (-80) = 64 + 80 = 144$

$3x^2 - x - 1 = 0$: $b^2 - 4ac = (-1)^2 - 4(3)(-1) = 1 - (-12) = 13$

Because $x^2 - 3x + 5 = 0$ is the only equation with a negative discriminant, (B) is correct.

8. B

The graph of a quadratic function is a parabola, a curve shaped like a U or an upside-down U. The range of the function is the set of possible y values of the coordinates of the points making up the curve. This includes the *vertex,* and all of the points above it (if the parabola is U-shaped) or below it (if the parabola has an upside-down U shape). To find the range of a quadratic function without graphing it, you can determine the coordinates of the vertex.

The vertex of the quadratic function $f(x) = a(x - h)^2 + k$ has the coordinates (h, k). To find the values of h and k for the function $f(x) = x^2 - 8x + 19$, we'll have to put it in the form more like $f(x) = a(x - h)^2 + k$.

If we square $(x-4)$, we get $x^2 - 8x + 16$. We can add 3 to that to get $x^2 - 8x + 19$. Therefore, $f(x) = a(x-h)^2 + k = f(x) = 1(x-4)^2 + 3$, and so the vertex is (4, 3). We know that the range is bounded by 3 on one side, then. Does the range include 3 and all the numbers greater than 3, or less than 3? That depends on the value of a in $f(x) = a(x-h)^2 + k$. If a is positive, the parabola is U-shaped; if not, it is upside-down U-shaped. Here, a is positive, and so the range of the function is (B), all real numbers greater than or equal to 3.

9. A

This question involves a longer polynomial function of the kind introduced in Chapter 4. Remember that a zero of a function $f(x)$ is a value of x for which $f(x) = 0$. You could simply test each of the numbers appearing in the answer choices and pick the choices that include all zeroes.

Another method involves factoring. We'll explain the division process you can use in Chapter 4. It would show that $x^3 + 3x^2 - 6x - 8 = (x+4) \bullet (x+1) \bullet (x-2)$. Since $x^3 + 3x^2 - 6x - 8 = 0$ if any of the factors equal 0, the zeroes are -4, -1, and 2. (A) is the answer.

10. A

We'll spend some time discussing domains of algebraic fractions or rational expressions in Chapter 4. Think of the domain of a function as the possible values x can have. If the denominator of a rational expression equals zero, then the expression is undefined. We can't have that, so any value of x that gives the expression an undefined value can't be part of the domain. So when would the denominator of $f(x) = \dfrac{x^3 + 8}{x^2 - 9}$ equal zero? To answer that question, we solve the equation $x^2 - 9 = 0$. The solutions are -3 and 3, (A). Any number other than those two is part of the domain, for the denominator has a value other than zero for any other value of x.

You might have picked (E) if you looked for a value of x that gives the numerator of the polynomial expression a value of 0. Remember that 0 is a value; having a value of 0 and being undefined are altogether different!

11. C

This question takes us to the next major topic: matrices. This is the subject of Chapter 5, and matrix addition is a very important part of it. As long as the two matrices to be added have the same dimensions, they can be added. Carrying out the addition is just a matter of adding the numbers in the corresponding parts of the matrix. For instance, we can begin by adding 3 and 0, since they are each located in the upper left-hand corners of their respective matrices. Since $3 + 0 = 3$, 3 will go in the upper left-hand corner of the matrix we get as a sum. We carry out the rest of the addition in the same way:

$$\begin{bmatrix} 3 & -4 & 5 \\ 5 & 2 & -7 \\ 8 & -8 & 1 \end{bmatrix} + \begin{bmatrix} 0 & 6 & 5 \\ 1 & -3 & 3 \\ -9 & -7 & -1 \end{bmatrix} = \begin{bmatrix} 3+0 & -4+6 & 5+5 \\ 5+1 & 2+(-3) & -7+3 \\ 8+(-9) & -8+(-7) & 1+-1 \end{bmatrix} =$$

$$\begin{bmatrix} 3 & 2 & 10 \\ 6 & -1 & -4 \\ -1 & -15 & 0 \end{bmatrix}$$

So (C) is correct. Be careful when pairing up corresponding numbers in the matrices. Also, be sure to handle negative numbers properly. The incorrect choices in this question all involve errors of these kinds.

12. A

Multiplying matrices is a little trickier than adding them. Keep in mind that two matrices don't need to have the same dimensions to be multiplied. There are certain conditions, though, and we'll review them in Chapter 5. To multiply matrices, you take each row in the first matrix and multiply it by each column in the second matrix. This involves multiplying the first number in a row of matrix 1 by the first number in a column of matrix 2, and the next number in that row of matrix 1 by the next number in the column in matrix 2, and so on. You add the products. That sum goes in one place in the product matrix. The process continues, and it can be long and complicated. We'll walk through it carefully later. For now, here is the multiplication process that gets you the answer:

$$\begin{bmatrix} 2 & 3 \end{bmatrix} \times \begin{bmatrix} 3 & -2 & 6 \\ 5 & -1 & -3 \end{bmatrix} =$$

$$\begin{bmatrix} (2(3) + 3(5)) & (2(-2) + 3(-1)) & (2(6) + 3(-3)) \end{bmatrix} =$$

$$\begin{bmatrix} (6 + 15) & (-4 - 3) & (12 - 9) \end{bmatrix} =$$

$$\begin{bmatrix} 21 & -7 & 3 \end{bmatrix}$$

The result is a 1 x 3 matrix

(D) is the result of multiplying the first matrix row by each row in the second matrix and putting the output in columns. (E) involves the same products but put in rows instead of columns.

13. E

Circles are one kind of conic section explored in Chapter 6. You can write the equation of a circle on a coordinate plane if you know the coordinates of the circle's center and the circle's radius. In this question, you can get that information by carefully examining the graph. The center of this circle is located at $(-3, 4)$. You can also see that the edge of the circle is 5 units from the center, whether you go left, right, down, or up. This means that the circle has a radius of 5. Now, the formula for a circle is $(x - h)^2 + (y - k)^2 = r^2$, where (h, k) is the center of the circle and r is the radius. So the equation is:

$$(x - (-3))^2 + (y - 4)^2 = 5^2, \text{ or } (x + 3)^2 + (y - 4)^2 = 25$$

(E) is the answer. (B) is close to the correct answer, but the radius is not squared there. (D) is the result of adding the coordinates of the center in the equation instead of subtracting them.

14. C

We can apply the circle formula introduced in the last problem to this new equation, $(x - 8)^2 + (y + 7)^2 = 49$. Here, $r^2 = 49$, so the radius of the circle is 7. The diameter of a circle is twice the radius. (C) is the correct answer. (A) is the value of the radius, and (E) is the square of the radius. The values of h and k in this equation, 8 and -7, don't even matter here.

15. C

Ellipses are another kind of conic section taken up in Chapter 6. There is a special formula we use to describe an ellipse with a horizontal major axis on the coordinate plane:

$$\frac{(x - h)^2}{a^2} + \frac{(y - k)^2}{b^2} = 1$$

In Chapter 6, we'll compare this formula to one for an ellipse with a vertical major axis. In this formula, (h, k) is the center of the ellipse, and a and b represent the lengths of the major and minor axes. The equation in the question doesn't include values for h and k. This means that they each have values of 0, and the ellipse is centered on the origin.

Since $a = 9$, the length of the major axis is 9, and (C) is the answer. Choice (A) gives the length of the minor axis. Choice (E) would be the result of taking a to represent the distance from the center to the edge of the ellipse along the major axis; doubling that number would get you 18.

16. D

Although there is no mention of logarithms in this question, you can use them to answer questions like these. Logarithms are the topic of Chapter 7. This question, like many other logarithm questions, requires either a scientific calculator or a table of logarithms. Now, in order to raise 3.5 to the power of 1.5, you first find the logarithm of 3.5. You then multiply the logarithm of 3.5 by 1.5, the power to which 3.5 is raised.

$\log 3.5 \approx 0.544068...$

$1.5 \cdot \log 3.5 \approx 1.5 \cdot 0.54407 \approx 0.816102...$

That last number is the logarithm of $3.5^{1.5}$. To get the value of the number, we find the *inverse logarithm* or *antilog* of 0.816102... You may have to look up the instructions on your calculator in order to do this. The antilog is 6.5479... Choice (D) is closest to that value.

17. E

As we'll see in Chapter 7, the laws of logarithms tell us that if $x^{1.5} = 343$, then $\log (x^{1.5}) = \log 343$. Also, $\log (x^{1.5}) = 1.5 \log x$. So $1.5 \log x = \log 343$.

$\log 343 \approx 2.53529...$

If we divide both sides of the equation by 1.5, we get

$\log x \approx 1.690196...$

Using a calculator to find the antilog of this number gives us a value of 49. So 49 raised to the power of 1.5 is 343, and (E) is the correct answer.

18. B

This is yet another question that can be dealt with the help of logarithms. Here, since $256^a = 1,048,576$,

$\log 256^a = \log 1,048,576$, and so $a \log 256 = \log 1,048,576$

Let's calculate the logarithms. The result is

$a(2.40823...) \approx 6.02059...$

Dividing both sides by the logarithm on the left side gets us

$a = 2.5$

We get 1,048,576 by raising 256 to the power of 2.5, and (B) is the correct choice.

19. C

This is a trigonometry question. That is the topic of Chapter 8. Sines, cosines, and tangents, among others, are trigonometric functions. They are functions of the measure of angles of right triangles. As it happens, the ratios of the sides of a right triangle determine the angle measures as well as the associated functions. The cosine of the angle of a right triangle (other than the right angle) is the ratio of the length of the adjacent side to the length of the hypotenuse. The side adjacent to angle A has a length of 36, and the hypotenuse, the side opposite the right angle, has a length of 39. So the cosine of A is $\frac{36}{39}$, or $\frac{12}{13}$, and choice (C) is correct. Choice (A) is actually the sine of the angle; the sine is the ratio of the opposite side (instead of the adjacent) to the hypotenuse. Choice (B) is the tangent of the angle; that function is the ratio of the opposite side to the adjacent side.

20. D

Once we have the sine, cosine, or tangent of an angle, we use the inverses of the trigonometric function to find the measure of the angle. For this, you need either a scientific calculator or a trigonometric function table. Let's find the sine of the angle measuring $x°$. Since we are given the measures of all three sides, we could just as readily use the cosine or tangent instead for our purposes, but we have to pick one of them. The sine is the ratio of the length of the side opposite the angle to the length of the hypotenuse. So the sine of our angle is $\frac{8}{10}$, or 0.8. Using the inverse sine function on the calculator, or a function table, we find that 0.8 is the sign of an angle measuring approximately 53.1°. (D) is the correct answer. Be careful in selecting the right inverse function. One who uses the inverse sine function on the cosine of the angle, 0.6, would find an angle measure of about 36.86°, which is close to choice (A).

21. B

Rotations are one kind of linear transformation that we'll study in Chapter 9. Rotations usually involve a figure being moved along an imaginary circle with the origin at its center. If you're standing on a merry-go-round in motion, not only does your body move, but the direction you are facing moves. Something similar goes on when figures are rotated about the origin. When figures are moved 180° in this way, the result is that each coordinate of the original point becomes its additive inverse. The additive inverse of a is $-a$. So the resulting coordinates here are (4, –8).

Let's visualize this. Starting with our original point, we move around the circle we describe. The point on the opposite side of the circle, 180° around the origin, has coordinates (–4, 8).

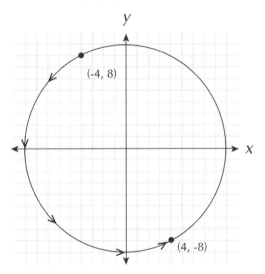

22. B

This question involves a different kind of transformation: a translation. You can think of translated figures as being "slid." They change position, but unlike rotated figures, they don't change direction. When a point with coordinates (a, b) is translated by (c, d), the new coordinates are $(a + c, b + d)$. So this triangle (or every point that makes it up), will be moved 6 units to the right and 2 units down. To identify the correct translation here, let's focus on the point that is the vertex of the right angle of the triangle. The vertex has coordinates (–3, 3). So the vertex of the translated triangle should have coordinates (3, 1). Only the triangle in the choice (B) has the vertex in that location. Choice (C) is the result of adding 2 to the y–coordinate of the original point, instead of subtracting.

23. B

Reflections are the other important kind of transformation discussed in Chapter 9. A reflection is just that: a figure reflected over a line on the coordinate plane. A very common type of question will ask you to identify the line of reflection. Notice that each point in the original figure has a corresponding point on the reflection. We could label the vertices of the figures to identify corresponding points.

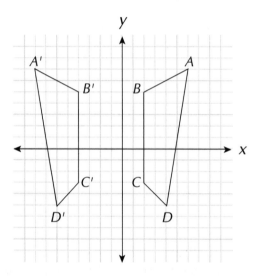

To identify a line of reflection, start with a pair of corresponding points. Let's take A and A'. We have to average their coordinates to find the midpoint of the line segment those points would form. A has coordinates (6, 7)' and A' has coordinates (–8, 7). So the midpoint is (–1, 7). The midpoint of the segment connecting B and B' is (–1, 5). This gives us enough information to answer the question. The line of reflection is the line that includes both midpoints. That is $x = -1$. So (B) is the correct answer. Choice (A) would be the result of getting 0 instead of –1 as the x–coordinates of the midpoints. (C) would be the result of getting –2 instead of –1 as the x–coordinates of the midpoints.

24. E

This question involves geometric sequences, one of the central topics of Chapter 10. A geometric sequence is a series of numbers that follows a certain pattern: the ratio of any number to the next one is the same throughout. Each number in the sequence is the product of the previous one and a fixed number. In the sequence given here, 8, 24, 72,..., that fixed number is 3; 24 is the product of 8 and 3, and 72 is the product of 24 and 3. So the next number in the sequence is the product of 72 and 3. That makes (E) the correct choice.

Choice (B) might be the result of identifying this as an arithmetic sequence instead of a geometric one. In an arithmetic sequence, the difference between any term and the next one is a fixed number. If you took the difference between 72 and 24, 48, and added that to 72, you would get 120.

25. B

As we just explained, in an arithmetic sequence, the difference between any number and the next one is a fixed amount. Here, we are given the third and fourth terms in a sequence and asked to work backward. Since the third term is 34 and the fourth is 41, we know that each term is 7 greater than the previous one. So the second term is 34 – 7, or 27, and the first term is 27 – 7, or 20. That makes (B) the correct answer. Choice (E) might be the result of taking 41 to be the third term and 34 to be the fourth term, instead of vice versa. That would make each term less than the previous one. You'll find that many arithmetic sequences follow such a pattern.

26. A

Many geometric series have an infinite sum. The term just keeps getting bigger, and the most we can do is find the sum of a limited part of the sequence. However, when each term is multiplied by a number greater than –1 and less than 1, the sum of the series will actually converge on a certain sum. The formula we'll use to find that number is $S = \dfrac{a}{1-r}$, where a is the first number in the series, r is the fixed number by which each term is multiplied to get the next one, and S is the sum of the series.

Here, $a = 36$ and $r = \dfrac{1}{3}$. So $S = \dfrac{36}{1-\frac{1}{3}} = \dfrac{36}{\frac{2}{3}} = \dfrac{3(36)}{2} = \dfrac{108}{2} = 54$

As the series goes on and on, the sum of the terms will get closer and closer to 54, without ever equaling 54 exactly. We can say that the sum of the infinite number of terms is actually 54, though.

27. C

This question brings us to the subject of permutations and combinations, which we'll take up again in Chapter 11. Here, we are asked to find the number of possible four-digit permutations of four digits. To answer the question, all we need to do is evaluate 4!. That expression is "4 factorial." A factorial expression is the product of the number given and every lesser integer, down to 1. So $4! = 4 \cdot 3 \cdot 2 \cdot 1 = 24$. So we can arrange the digits 2, 3, 4, 5 in 24 different ways, so as to make 24 different numbers, including 2,345, 5,342, etc. Note that this total would be the same no matter what four digits we were given. There are 24 permutations of the digits, 1, 6, 7, and 9, for instance.

28. A

This question really involves combinations, as opposed to permutations. Here, we are trying to find out how many combinations of three people we can make from a larger group of five. This question also differs from the previous one because here, the order in which the people are combined doesn't matter.

We'll apply a more complicated formula, and then explain fully why it works in Chapter 11.

The number of possible combinations is given by the formula

$C = \dfrac{n!}{m!(n - m!)}$, when n is the size of the larger group, and m is the size of the smaller group.

Here, $m = 3$ and $n = 5$, so $C = \dfrac{5!}{3!(5 - 3!)} = \dfrac{5!}{3!(2!)} = \dfrac{120}{6(2)} = \dfrac{120}{12} = 10$.

So ten possible committees can be formed. We couldn't just go with 5!, which is the value of choice (E), because the committee is not 5 members large. We divide 5! by 2! to account for the smaller committee size. We also divided by 3! to account for the factor that the order of the selection doesn't matter. The value of $\frac{5!}{2!}$ would give us the number of "ordered" committees. If order matters, then a committee of Adam, Brenda, and Chris is different from the committee of Brenda, Chris, and Adam. Since we treat them as the same, and there is actually just one unordered committee for every six ordered ones, we divide by 3!, or 6. You would have gotten choice (D) if you hadn't done that.

29. C

The final chapter of this book covers probability. It will take up compound probability, just as this question does. It involves compound probability because it deals with a series of events. Here, the events are "independent." This means that the outcome of one event does not affect the likelihoods of the outcomes of another event. To find the probability of several independent events all occurring, you find the product of the probability of each one occurring. The probability of a fair coin landing on heads after a toss is 50%, 0.5, or $\frac{1}{2}$. So the probability of the coin landing on heads all three times is $\frac{1}{2} \cdot \frac{1}{2} \cdot \frac{1}{2} = \frac{1}{8}$.

To better understand why the probability is $\frac{1}{8}$, look at this table.

	First Flip	Second Flip	Third Flip
1	Heads	Heads	Heads
2	Heads	Heads	Tails
3	Heads	Tails	Heads
4	Heads	Tails	Tails
5	Tails	Heads	Heads
6	Tails	Heads	Tails
7	Tails	Tails	Heads
8	Tails	Tails	Tails

The table gives every possible outcome of the series of three coin tosses. There are eight altogether. Out of these eight outcomes, only one of them, the first one, has what we are looking for. This probability can also be understood, then, as the ratio of the number of desired outcomes to the number of possible outcomes, which is $\frac{1}{8}$.

30. A

This probability question doesn't involve separate, independent events as the previous question does. Rather, it involves an event with a number of desired outcomes. That's because there are several ways you can roll a total 10. First of all, there are 36 possible outcomes of a roll of a pair of dice. You could roll 1s with each die; you could roll 1 with the first die, and 2 with the second, or vice versa. Here are all the possible rolls and totals:

	1	2	3	4	5	6
1	2	3	4	5	6	7
2	3	4	5	6	7	8
3	4	5	6	7	8	9
4	5	6	7	8	9	**10**
5	6	7	8	9	**10**	11
6	7	8	9	**10**	11	12

The row headings represent the number you get with the first die. The column headings represent the number you get with the second die. We highlighted the rolls that get a total of 10. Out of 36 possible roll totals, 3 of them get a total of 10. So the probability of getting that total is $\frac{3}{36}$, or $\frac{1}{12}$.

Thus, (A) is the answer. You might have gotten (C) if you counted 4 ways instead of 3 of getting a total of 10.

DIAGNOSTIC TEST CHART

Question Number	Topic	Chapter in Which the Topic Is Covered
1	Evaluating Functions	1
2	Finding a Function's Range	1
3	Finding the Zero of a Function	1
4	Intercepts of Linear Functions	2
5	Identifying Linear Functions	2
6	Finding the Zeroes of a Quadratic Function	3
7	Quadratic Function with Imaginary Roots	3
8	Finding a Quadratic Function's Range	3
9	Finding the Zeroes of a Polynomial Function	4
10	Finding the Domain of a Rational Function	4
11	Adding Matrices	5
12	Multiplying Matrices	5
13	Finding the Equation of a Circle	6
14	Finding the Properties of a Circle	6
15	Finding the Properties of an Ellipse	6
16	Using Logarithms to Evaluate Powers	7
17	Using Logarithms to Find Bases	7
18	Using Logarithms to Find Exponents	7
19	Evaluating Trigonometric Functions	8
20	Using Trigonometric Functions to Find Angle Measures	8

Question Number	Topic	Chapter in Which the Topic Is Covered
21	Rotating Points	9
22	Translating Figures	9
23	Identifying Lines of Reflection	9
24	Continuing Geometric Sequences	10
25	Identifying Terms in Arithmetic Sequences	10
26	Finding the Sum of an Infinite Geometric Series	10
27	Calculating the Number of Permutations	11
28	Calculating the Number of Combinations	11
29	Compound Probability (Independent Events)	12
30	Simple Probability	12

CHAPTER 1

Functions

WHAT ARE FUNCTIONS?

The function is one of the most important concepts in Algebra II. This chapter is just a basic introduction to functions; we'll be dealing with them throughout the book.

Functions are special kinds of relations. A function takes a given set of values, and relates each of them to another value. We'll call these values *elements*. The first set of elements is called the *domain*. Each element in the domain is paired with an element in another set, known as the *range*.

Many functions are presented as equations. In fact, much of basic algebra involves functions. Many linear equations, for instance, are functions; they're just not always identified with that term. In fact, the equation $y = 2x$ is a function. We'll explain how we can write this function with different notation.

In this chapter, we will explain how to identify functions, as well as their domains and ranges. We will also go through basic aspects of evaluation and equation solving.

CONCEPTS TO HELP YOU

We are about to introduce some of the most basic function concepts. It is very important to understand these, as they will come up often throughout this book. The concepts of domain, range, and zeroes, to name a few, are developed in later chapters covering functions.

i. Functions and Relations

All functions are relations, but not all relations are functions. A relation is really just a set of ordered pairs, a way of relating information. If you take a group of people, a relation might relate each member of the group to some value, such as age. For a group of four people, a relation would organize that age information in terms of ordered pairs:

(Adam, 14)
(Ben, 16)
(Carla, 15)
(Diane, 17)

This relation tells us that Adam's age is 14, Ben's is 16, and so on. The domain of this relation is the four people: Adam, Ben, Carla, and Diane. The range is the ages: 14, 15, 16, and 17.

This relation also happens to be a function. That's because each member of the domain is paired with exactly one member of the range. Not all relations are functions, however, because some relations pair a member of the domain with more than one member of the range. One good example involves square roots. You might have a relation that pairs positive integers with their square roots. It would include the following ordered pairs, among others:

(9, 3)
(9, –3)
(4, 2)
(4, –2)

Recall that numbers such as 4 and 9 have positive and negative square roots. So 9, a member of the domain, is paired with two different members of the range, 3 and –3. That's why this relation is *not* a function.

On the other hand, there is a relation that pairs integers with just their positive square roots. Each positive integer has exactly one positive square root, so that relation is a function.

We've been talking about relations and functions in terms of ordered pairs, but they are not always presented as ordered pairs. Many functions are represented as equations. We mentioned the function $y = 2x$ earlier. There, the domain is the set of values x could have, and the range is the set of values of y. Note that the value of y is dependent on the value of x. We take an element of the domain and match it up with the element of the range that has twice the value. So we could use the equation to come up with a long list of ordered pairs, including (1, 2), (2, 4), (10, 20), (24, 48), and so on.

ii. Function Notation

An equation that relates each member of a domain to a member of a range is a function. The equation $y = 2x$ is a function. In fact, many of the algebraic equations you've already encountered are functions. In Algebra II, you'll frequently see the notation $f(x)$, instead of the letter y. This would be read aloud as "f of x." It doesn't involve multiplying a variable f by a variable x, although it might look that way. When you see the statement $f(x) = 2x$, a function of x is defined as $2x$. Sometimes functions use the letter g instead of f. The variable x is called the *argument*.

So why bother replacing y with $f(x)$? There are a number of reasons. Some algebra problems involve multiple functions. It would be hard to keep track of them all with the same group of variables. Instead, we could use different letters for different functions. Rather than dealing with a pair of functions written as $y = 2x$ and $y = x + 5$, we could handle them as $f(x) = 2x$ and $g(x) = x + 5$.

Also, this new function notation makes certain algebraic tasks easier to state. Instead of asking "what is the value of y if $y = 2x$ and $x = 4$?", we can ask "what is $f(4)$ if $f(x) = 2x$?"

LETTERS AND FUNCTIONS

The most common letters used to label functions are f and g. Common as well are h, j, and k. Letters that are often used as arguments, such as x, y, and z, are not often used as labels.

The variable x is the most common argument you'll find in algebraic functions, but any variable is fair game. Don't be thrown off by functions such as $f(a) = a^2 + 4$ or $f(d) = -7d$. It doesn't matter which letter is used, as long as it is used consistently.

Some functions have more than one argument. The equation $f(a, b) = 3a + 4b$ has two arguments, a and b. Our focus in this book, though, will be functions with a single argument.

iii. Domain

You can think of the domain of a function as the set of possible values for the argument of a function. For many functions, the domain is the entire set of *real numbers*, which consists of all of the numbers on a number line. Any rational or irrational number can be a value.

Many functions, however, have limited domains. One requirement is that the value of the function be a real number as well. Take the $f(x) = \sqrt{x}$ Since $\sqrt{-1}$ is not a real number, -1 cannot be a part of the domain of $f(x)$. No negative number is part of that domain, for that matter.

In this case, we could say that the function's domain is $x \geq 0$, or all real numbers greater than or equal to zero.

In some cases, a function might have a specified domain. A question might say that $g(x) = 4x - 6$, where $5 \leq x \leq 10$. That inequality says that the domain of the function is the set including 5 and 10, and all real numbers between them.

iv. Range

Even if the domain of a function is the set of all real numbers, the range of the function might not be that set. The range is the set of possible values of a function. Suppose that $f(x) = x^2$. While that argument x can have any real number value, $f(x)$ cannot be less than 0. That's because $x^2 \geq 0$ for any real number x. So the range of the function is $f(x) \geq 0$, or all real numbers greater than or equal to zero.

v. Evaluating Functions

Evaluating a function is very much like evaluating an algebraic expression. Instead of asking you to evaluate, say, $x + 10$, when $x = -3$, an Algebra II question might ask for the value of $f(-3)$ if $f(x) = x + 10$. Evaluating the function is just a matter of plugging the argument -3 into the expression, and simplifying. So $f(-3) = -3 + 10 = 7$.

Sometimes, one function operates on another. When that happens, you have a *compound function*. You might encounter a question that defines $f(x)$ and $g(x)$, and asks you to evaluate $f(g(a))$, where a is a real number. As you'll see below, the key to dealing with a compound function is to treat the function inside the parentheses as the argument for the one outside.

vi. Zeroes

Finding the zero of a function involves equation solving. It's important to know, first of all, that finding the zero of a function $f(x)$ is not the same thing as evaluating $f(0)$. Rather, you find the zero of $f(x)$ by finding the value of x where $f(x) = 0$.

Take the function $f(x) = x - 5$. Since $x - 5 = 0$, if $x = 5$, 5 is the zero of that function.

Some functions have more than one zero. Many quadratic functions have two zeroes, as we'll see in Chapter 3. Other polynomial functions can have even more, as we'll see in Chapter 4.

STEPS YOU NEED TO REMEMBER

We will go through many of the key steps you'll need to take to set up solutions to function questions. As we've said, we'll continue to explore functions in many of the upcoming chapters. Many of these basic steps will apply well beyond this chapter.

i. Identifying Functions

To determine whether a set of ordered pairs is a function, you need to identify each element of the domain. See whether any elements of the domain appear in more than one ordered pair. Suppose you find an element that appears twice. Is it paired with the same element of the range both times? If not, then the relation is not a function. If no element of the domain is ever paired with two different elements of the range, then the relation is a function.

The steps just discussed amount to the *vertical line test*. You can plot a set of ordered pairs on the coordinate plane, using the element of the domain as the *x*-coordinate of each point, and the element of the range as the *y*-coordinate of each point.

Let's put the function {(4, 3) (5, 2), (6, 1), (5, 4)} to the test. Plot each of those points on the plane, and draw a vertical line through each:

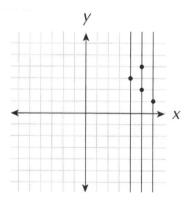

One of the lines passes through two points. This indicates that a number in the domain is paired with two different numbers in the range. Here, 5 is paired with both 2 and 4. The relation fails the vertical line test, then, and is not a function. A relation is a function only if no vertical line passes through more than one point.

ii. Identifying the Domain

Every number in the domain of a function must give the function a real number value. Any number that doesn't is not a part of the domain. Many of the function types we'll introduce in this book have their own particular domain restrictions. We can't get into all of them here, so we'll focus on a couple of situations.

Suppose a function has a term that must be greater than or equal to zero. One such term is a radical: $\sqrt{x}$ not a real number if $x < 0$. If you have a longer expression inside the radical, then you must solve an inequality for the variable. For instance, if $f(x) = \sqrt{x+1}$, then $x + 1 \geq 0$. You have to solve the inequality to find the domain of the function. Here, $x \geq -1$, so that is the domain of $f(x) = \sqrt{x+1}$

Another restriction on the domain of a function has to do with denominators. An algebraic fraction is undefined if the denominator has a value of 0. So the variable cannot have a value that makes the denominator 0. Take the $f(x) = \dfrac{1}{2x-1}$. The function is undefined if $2x - 1 = 0$. To see what value of x can't be part of the domain, solve that equation. The solution $x = \dfrac{1}{2}$, so $\dfrac{1}{2}$ is not part of the domain. In fact, the domain is all real numbers other $\dfrac{1}{2}$.

iii. Identifying the Range

To identify the range of a function, you should first identify any terms that have a limited range of values. Use those in an inequality. If part of the function must have a value greater than 0, for instance, state that as part of an inequality. Then perform operations on both sides of the inequality to get the function on one side. If you are able to get the function on one side and a real number on the other, then you will have identified the range.

Take the function $f(x) = |x| + 2$. Since an absolute value cannot be negative, $|x| \geq 0$. Since the function is $|x| + 2$, add 2 to both sides of the inequality to get the function on one side. The result is $|x| + 2 \geq 2$. So the range is $f(x) \geq 2$.

There are special methods for determining the ranges of certain kinds of functions. We'll discuss how to find the range of a quadratic function, for instance, later in the book.

iv. Evaluating Functions

To evaluate a function $f(a)$, where a is the real number argument, replace the variable in the function with the number, and simplify.

When you have a compound function, you need to apply the functions in the right order. Given functions $f(x)$ and $g(x)$, you would evaluate $f(g(5))$ by first evaluating $g(5)$. Then evaluate $f(x)$ for the value you find for $g(5)$. You always evaluate the innermost function first, and then work your way outward.

v. Finding the Zero

As explained above, finding the zero of a function is a matter of equation solving. Take the function $f(x) = A$, where A stands for some algebraic expression with the variable x. You can find the zero by replacing $f(x)$ with 0 to get the equation $0 = A$. Solve this equation for x. The solution is the zero of the function.

STEP-BY-STEP ILLUSTRATION OF THE FIVE MOST COMMON QUESTION TYPES

Now that we've reviewed the fundamental concepts and problem-solving steps, it is time to apply them. We'll walk through the process of solving five different kinds of basic function questions. Keep in mind that there are many kinds of functions. We can't work with that many of them here, but we can display strategies you can apply to other functions.

Question 1: Identifying Functions

Which of the following sets of ordered pairs could be a function?

(A) (–5, 1), (–3, 8), (–5, 0), (–3, 5)

(B) (–4, –4), (6, 3), (4, –3), (–4 , 0)

(C) (0, 9), (6, 0), (0, 0), (3, 9)

(D) (–1, –4), (2, 5), (5, –5), (6, –4)

(E) (1, 2), (1, 3), (2, 5), (5, 6)

In a function, each of the first members of the ordered pairs is matched with a single value. In other words, each member of the domain is paired with a single number in the range. So, if the same number, appearing first in ordered pairs, is paired with two different numbers, then the set of ordered pairs is *not* a function.

Let's go through each set of ordered pairs, then, to see which passes the test. Choice (A) includes the ordered pairs (–5, 1) and (–5, 0). Since –5 is paired with two different numbers, this set fails the test and is not a function. Choice (B) can also be ruled out because it includes (–4, –4) and (–4, 0); –4 is paired with two different numbers. Choice (C) pairs 0 with 9 in the ordered pair (0, 9) and 0 in the pair (0, 0). Choice (E) pairs 1 with 2 in the first pair and 3 in the second pair. Therefore, none of those choices represent functions.

Choice (D) never pairs a number appearing first with two distinct numbers. It's the one set of pairs that could be a function. **So, (D) is the correct answer choice.** It doesn't matter that –4, appearing twice as the second number in ordered pairs, is paired with different numbers. That's because –4 appears there as part of the range. Again, the only requirement is that each element in the domain is paired with only one element in the range.

Question 2: Identifying the Domain

What is the domain of the $f(x) = \sqrt{5-x}$?

(A) $x \geq -5$

(B) $x \leq 0$

(C) $x \geq 0$

(D) $x \leq 5$

(E) $x \geq 5$

In order for $f(x)$ to be a real number, the value of $\sqrt{5-x}$ must be a real number. However, $\sqrt{5-x}$ does not have a real number value if $5 - x$ is less than zero. You cannot have a radical of a negative number. So, $5 - x$ must be greater than or equal to zero. You can determine the domain of this function, then, by finding the values x can have. This is a matter of solving the $5 - x \geq 0$ unequality:

$$
\begin{array}{r}
5 - x \geq 0 \\
\underline{-5 \qquad -5} \\
-x \geq -5 \\
\underline{x-1 \quad x-1} \\
x \leq 5
\end{array}
$$

So, choice (D) is the correct answer. Remember that when you multiply both sides of an inequality by a negative number, the direction of the inequality sign is reversed. If you overlooked that step, you would have gotten choice (E). Choice (C) would be the correct answer if the question had asked for the range, rather than the domain. Therefore, $f(x)$ can be any real number greater than or equal to zero. But x, on the other hand, can be less than zero; it just can't be greater than 5.

Question 3: Identifying the Range

What is the range of the function $g(x) = |x + 3| - 4$?

(A) $g(x) \leq -4$

(B) $g(x) \geq -4$

(C) $g(x) \geq -1$

(D) $g(x) \leq 3$

(E) $g(x) \geq 3$

The range of the function $g(x)$ is the set of the possible values $|x + 3| - 4$. Since this function involves absolute values, you should keep in mind that the absolute value expression must have a value of 0 or higher; absolute values are never negative.

When $1x + 31 = 0$, $g(x) = 1x + 31 - 4 = -4$. That is the lowest possible value of $|x + 3| - 4$ since $|x + 3|$ cannot be less than zero. On the other hand, there is no maximum value of $|x + 3| - 4$. Since the domain of the function is all real numbers, x can have any value. This is an absolute value function, and so its value increases as the absolute value of x increases. It can be any number greater than or equal to -4, **so (B) is the correct answer choice.** Choice (A), $g(x) \leq -4$, is the range of the function $g(x) = -4 - |x + 3|$. Since $|x + 3|$ cannot be less than zero, that function cannot be greater than -4. Choice (C), $g(x) \geq -1$, is the range of the function $g(x) = |x| + 3 - 4$. Choice (D) is the range of the function $g(x) = 3 - |x + 4|$ Choice (E) is the range of the function $g(x) = |x + -4| + 3$.

Question 4: Evaluating Functions

If $f(x) = \sqrt{11x + 5}$, then $f(4) =$

(A) 7

(B) 16

(C) 27

(D) 33

(E) 77

Finding the value of $f(4)$ is just a matter of evaluating $\sqrt{11x + 5}$ when $x = 4$. This means that you substitute 4 for x in the expression and simplify. If $x = 4$, then

$$\sqrt{11x + 5} =$$
$$\sqrt{11(4) + 5} =$$
$$\sqrt{44 + 5} =$$
$$\sqrt{49} = 7$$

So, $f(4) = 7$, and **choice (A) is the correct answer.** Choice (C), 27, is actually the value of $11\sqrt{x} + 5$ when $x = 4$. Choice (D), 33, is the value of $11\sqrt{x + 5}$ when $x = 4$. Choice (E), 77, is the value of $11(\sqrt{x} + 5)$ when $x = 4$.

Question 5: Finding the Zero

What is the zero of $f(x) = 18 - 2x$?

(A) −18

(B) −9

(C) −2

(D) 2

(E) 9

The zero of this function is the value of x for which $f(x) = 0$. Since $f(x) = 18 - 2x$, you need to solve the equation $18 - 2x = 0$. That's just a matter of straightforward algebra:

$$
\begin{array}{r}
18 - 2x = 0 \\
\underline{-18 \qquad -18} \\
-2x = -18 \\
\underline{\div -2 \quad \div -2} \\
x = 9
\end{array}
$$

So choice (E) is the correct answer. Choice (A), −18, is the value of −2x. If $x = -9$, as in choice (B), $f(x) = 18 - 2(-9) = 18 + 18 = 36$. You might get that on one side of the equation in the next-to-last step, so be sure to solve the equation completely. While choice (C), −2, is the number by which you multiply the zero in the function, it's not the function itself. If $x = 2$, as in choice (D), $f(x) = 18 - 2(2) = 18 - 4 = 14$.

CHAPTER QUIZ

1. Which of the following sets of ordered pairs cannot be a function?

    (A) {(2, 4), (3, 4), (5, 4)}

    (B) {(4, 7), (3, 6), (3, 9)}

    (C) {(5, 1), (4, 2), (7, 3)}

    (D) {(6, 3), (7, 8), (9, 9)}

    (E) {(8, 7), (9, 4), (11, 3)}

2. A set of ordered pairs is plotted on the coordinate grid below.

    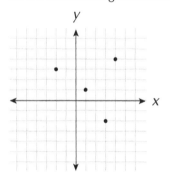

    A set of ordered pairs including the ones plotted above cannot be a function if it also includes which of the following pairs?

    (A) (–3, 3)

    (B) (–1, 1)

    (C) (2, 1)

    (D) (3, 0)

    (E) (5, 4)

3. What is the domain of the function {(1, 5), (4, 6)}?

    (A) {1, 4}

    (B) {1, 5}

    (C) {1, 6}

    (D) {4, 6}

    (E) {5, 6}

4. What is the domain of $f(x) = x^2 - 3$?

    (A) All real numbers

    (B) All real numbers less than –3

    (C) All real numbers less than or equal to 3

    (D) All real numbers greater than or equal to –3

    (E) All real numbers greater than 3

5. Which of the following numbers is not part of the domain $f(x) = \dfrac{3x - 12}{4x - 20}$?

    (A) 2

    (B) 3

    (C) 4

    (D) 5

    (E) 6

6. The ordered pairs of a function are plotted on the coordinate grid below.

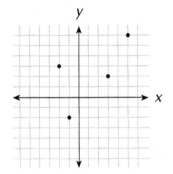

What is the range of the function?

(A) {–3, –1, 2, 5}

(B) {–2, –1, 3, 5}

(C) {–2, 2, 3, 6}

(D) {–1, 2, 3, 6}

(E) {1, 2, 3, 5}

7. Which of the following functions has a range of $f(x) \le 4$?

(A) $f(x) = \sqrt{x-4}$

(B) $f(x) = \sqrt{x} - 4$

(C) $f(x) = \sqrt{x+4}$

(D) $f(x) = \sqrt{x} + 4$

(E) $f(x) = 4 - \sqrt{x}$

8. If $f(x) = 4x^2 - 3x$, then what is the value of $f(-5)$?

(A) –415

(B) –115

(C) 85

(D) 115

(E) 415

9. $g(x) = 4x + 3$ and $h(x) = x^2 + 1$. $g(h(-2)) =$

(A) 17

(B) 20

(C) 23

(D) 26

(E) 29

10. What are the zeroes of $f(x) = |x - 6| - 5$?

(A) –11 and 11

(B) –1 and –11

(C) –1 and 11

(D) 1 and –11

(E) 1 and 11

ANSWER EXPLANATIONS

1. B

In the set $\{(4, 7), (3, 6), (3, 9)\}$, 3 is paired with both 6 and 9. Since a number in the domain is paired with two numbers in the range, the relation is not a function. Choice (A) might be tempting because there is only one number in the range. That doesn't matter though. A function can pair a single number in the range with different numbers in the domain.

2. D

The ordered pairs plotted on the grid are $(-2, 3)$, $(1, 1)$, $(3, -2)$, and $(4, 4)$. So the elements of the domain represented there are -2, 1, 3, and 4. Now the pair, $(3, 0)$ includes one of those elements, but pairs it with a different number in the range. Plotting that ordered pair shows that a relation including $(3, 0)$ cannot be a function.

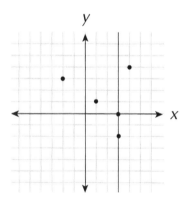

3. A

This function consists of just two ordered pairs. The domain is the set including the first number in each pair. The first number in the first pair is 1, and the first number in the second pair is 4. So the domain is $\{1, 4\}$. Choice (B), $\{1, 5\}$, is just the first ordered pairs, which includes one element from the domain and one from the range. Choice (E), $\{5, 6\}$, is the range.

4. A

There are actually no restrictions on the domain of this function. No matter the value of x, $x^2 - 3$ has a real number value. Choice (D) actually gives the range of the function, rather than the domain.

5. D

The denominator of an algebraic fraction cannot have a value of 0. Here, it would have that value if $4x - 20 = 0$. Since $x = 5$ if $4x - 20 = 0$, 5 cannot be part of the domain. Choice (C), 4, is the solution to $3x - 12 = 0$. Keep in mind that the numerator of a fraction can equal zero. Having a value of zero is not the same as being undefined, so 4 is part of the domain.

6. C

The coordinates of the points are $(-2, 3)$, $(-1, -2)$, $(3, 2)$, and $(5, 6)$. The range consists of the second number in each ordered pair: 3, -2, 2, and 6. Choice (B) is actually the domain of the function.

7. E

The range of a function includes only defined values, and a radical expression is undefined if the value inside the radical sign is less than zero. Since a radical is a positive square root, a radical with a real number value must be greater than or equal to zero. Now, the difference between four and any number greater than or equal to zero must be less than four. So, the range of $f(x) = 4 - \sqrt{x}$ includes no number greater than four.

Let's look at the other choices. $f(x) = \sqrt{x - 4} = 5$ if $x = 29$. $f(x) = \sqrt{x} - 4 = 5$ if $x = 81$. $f(x) = \sqrt{x + 4} = 5$ if $x = 21$. $f(x) = \sqrt{x} + 4 = 5$ if $x = 1$. So, all of the other functions have ranges including a number greater than four.

8. D

Since $f(x) = 4x^2 - 3x$, $f(-5) = 4(-5)^2 - 3(-5) = 4(25) - (-15) = 100 + 115$ $= 115$. Choice (C), 85, is the value of $f(5)$. Choice (E), 415, is the value of $f(-5)$ when $f(x) = (4x)^2 - 3x$ instead of $f(x) = 4x^2 - 3x$.

9. C

$h(-2) = (-2)^2 + 1 = 4 + 1 = 5$. So $g(h(-2)) = g(5) = 4(5) + 3 = 20 + 3 = 23$. Choice (D) is actually the value of $h(g(-2))$. Be careful to evaluate the functions in the right order.

10. E

This question asks for the plural "zeroes"; that's because many absolute value equations have more than one solution. Since you need to find out $|x - 6| - 5 = 0$, you actually have to solve two equations: $(x - 6) - 5 = 0$, in the event that $x - 6$ is positive, and $-(x - 6) - 5 = 0$, in the event that $x - 6$ is negative.

If $(x - 6) - 5 = 0$, then $x - 11 = 0$, and $x = 11$. If $-(x - 6) - 5 = 0$, then $-x + 6$ $- 5 = 0$, and $x = 1$. So, the zeroes of $f(x)$ are 1 and 11. Choice (A) is the result of getting $-x - 6 - 5 = 0$ instead of $-x + 6 - 5 = 0$ when solving $-(x - 6) - 5$ $= 0$.

Linear Functions

WHAT ARE LINEAR FUNCTIONS?

A linear function is one whose graph is a straight line. We are interested in linear functions mainly because the properties of straight lines are important, and because those are useful in representing many real-world situations. In particular, the properties of *slope* and *intercepts* need our attention. By using the properties of linear functions, we can graph them without having to plot many individual points.

CONCEPTS TO HELP YOU

The properties of lines and linear functions are useful for a number of purposes. We can use them to graph lines, and we can recognize particular functions from their graphs. You will find that knowing the coordinates of two points on a graph, or the slope of the graph and one set of coordinates, is enough to define a linear function.

i. Linear Relations and Functions

A linear relation is a set of ordered pairs whose graph is a straight line. Not all linear relations are functions, though. Look at this graph.

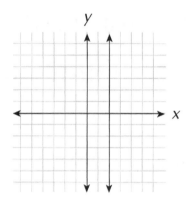

This is the graph of $x = 2$. If x is supposed to be the domain in this relation, such that it is made up of the ordered pairs (2, 3), (2, 4), (2, –1), and so on, then it is not a function. The number in the domain, 2, is paired with more than one member of the range. Any line other than a vertical one is not just a linear relation, but also a linear function.

Unlike many other algebraic functions, linear functions have domains and ranges that include all of the real numbers, unless otherwise specified. You can plug any real number into a linear function to get a unique member of the range.

ii. Slope

The slope of a line is a number that describes its direction and "steepness." The slope is the ratio of the difference in y-coordinates of two points on the line to the difference between their x-coordinates. The letter m is commonly used as the variable representing slope in linear functions. If you take points on the line with coordinates (x_1, y_1) and (x_2, y_2), then the slope is

$$m = \frac{y_2 - y_1}{x_2 - x_1}$$

So, if a line has a slope of 2, it means that for every unit increase for x, the value of the y-coordinate increases by 2. Likewise, if a line has a slope of 4, then for every unit increase for x, the value of the y-coordinate increases by 4. That is, the y-coordinates increase at four times the rate of the x-coordinates.

That relationship can be expressed with an equation. If $f(x) = 4x$, then the value of $f(x)$ increases four times the rate of increase of x. So, the function $f(x) = mx$, describes a line with slope m.

The x-intercept of that function is $f(0)$. There, $f(0) = 0$. But what if we have a linear function where $f(0) \neq 0$? Suppose $f(0) = b$, where b is some real number. Then the y-intercept would have coordinates $(0, b)$, and the function would be $f(x) = mx + b$.

This linear function $f(x) = mx + b$ appears in a very common and useful form: the *slope-intercept form*. There, once again, m is the slope and b is the y-intercept of the line.

iii. Intercepts

A straight line on the coordinate plane crosses at least one axis. Unless it is a perfectly horizontal or vertical line, it crosses both axes. The points where a line crosses the axes are called the *intercepts*. The x-intercept of a line always has a y-coordinate of 0, and the y-intercept of a line always has an x-coordinate of 0. Since one coordinate of each intercept is already known, we can usually identify an intercept with a single number, rather than an ordered pair. For instance, if a line passes through the point $(7, 0)$, we could say that the line has an x-intercept of 7.

Linear functions can be used to identify the intercepts of a line. In that case, we can graph the equation of a line by plotting a straight line through both intercepts. In fact, all you ever need to graph a line is two points. Suppose we know that a line has an x-intercept of -3 and a y-intercept of 4. You can plot the two points, and draw a line that passes through both of them.

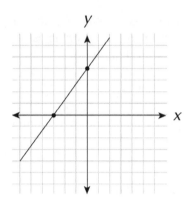

There is only one line that passes through both intercepts.

iv. Using the Slope-Intercept Form

Once you have a linear function in slope-intercept form, you can read the y-intercept right off the equation. Getting the x-intercept will require just a little more work, however.

Let's take the equation we looked at the previous section $y = \dfrac{3x}{2} + 4$. In this equation of the form $y = mx + b$, $b = 4$, and so the y-intercept is 4. Now, to find the x-intercept, you need to solve the equation for x when $y = 0$. The idea is that you need to find the x-coordinate of the point on the line where the y-coordinate is 0. So, if $y = 0$,

$$\dfrac{3x}{2} + 4 = 0$$

Let's solve this equation for x. First, subtract 4 from both sides to get

$$\dfrac{3x}{2} = -4$$

Now, since the variable x is multiplied by $\dfrac{3}{2}$ you need to divide both sides by that number. That amounts to multiplying both sides by the multiplicative inverse $\dfrac{2}{3}$.

$$\frac{3x}{2} \cdot \frac{2}{3} = -4 \cdot \frac{2}{3}$$

$$x = -\frac{8}{3}$$

And so the *x*-intercept of the line is $-\frac{8}{3}$

Once you have both intercepts of a line, you can graph the equation by plotting those two points and drawing a line through them.

There is a shortcut to the process for finding the *x*-intercept. Once you have the linear equation in the form $y = mx + b$, the *x*–intercept is $-\frac{b}{m}$. Notice that we actually just solved the equation for *x* when $y = 0$ by dividing $-b$ by m.

STEPS YOU NEED TO REMEMBER

You should be ready to deal with linear functions in whichever form they appear. They can appear as equations, tables, sets of ordered pairs, or graphs. Many questions will ask you to relate these forms somehow. It is important that you are ready to represent linear graphs in algebraic terms, linear functions in tables, and so on.

i. Recognizing the Features of Linear Functions

A linear function either takes the form $f(x) = mx + b$ or can be rewritten in that form. Any function including a simplified expression where the variable is divided, raised to a power, or put inside a radical or root sign is not a linear function.

It is important to remember that *m* or *b* can equal 0. The function $f(x) = 4x$ is a linear function where $b = 0$. Likewise, the $f(x) = 4$ is a linear function where $m = 0$.

ii. Using Tables and Ordered Pairs

A common question type will ask you whether a table, containing elements of the domain paired with elements of the range, could represent a linear function. It might instead ask what particular linear function the table represents. Answering that last question usually calls for a process of trial and error.

Suppose you are asked whether the table below represents the function $f(x) = 2x + 5$:

x	$f(x)$
4	13
6	19

To answer this question, you have to see whether $f(x) = 2x + 5$ for each row in the table. To do that, evaluate the function for each value of x:

$f(4) = 2(4) + 5 = 8 + 5 = 13$

$f(6) = 2(6) + 5 = 12 + 5 = 17$

The value of $f(4)$ matches the value of $f(x)$ in the first row, but the value of $f(6)$ does not match the second row. So, the answer is that the table does not represent $f(x) = 2x + 5$. Every row of the table has to match the function in question.

iii. Using the Coordinate Plane

Many questions in this area of Algebra II will require you to identify points on the coordinate plane. Remember always that the first digit of a pair of coordinates is the x-coordinate and the second digit is the y-coordinate.

Let's go through some of the strategies you should be ready to use to answer common questions.

The following are steps you can take to use a linear function to graph a line:

- Put the function in standard form, if it is not already so.
- Find the intercepts. The *x*-intercept is $-\dfrac{b}{m}$ and the *y*-intercept is *b*.
- Plot the intercepts and draw a line through them.

or

- Put the function in standard form, if it is not already so.
- Evaluate the function for two values of *x*. This will give you two pairs of coordinates.
- Plot both points and draw a line through them.

The following are steps you can take to define a linear function from its graph:

- Identify two pairs of coordinates.
- Use those coordinates in the slope $m = \dfrac{y_2 - y_1}{x_2 - x_1}$
- Plug the slope and a pair of coordinates into the equation $y = mx + b$. Solve the equation for *b* to get the *y*-intercept.

STEP-BY-STEP ILLUSTRATION OF THE FIVE MOST COMMON QUESTION TYPES

We have now reviewed the key concepts and steps you'll need to know to answer linear function questions. Now we'll apply them to the kinds of questions you're likely to encounter in the future. Below are questions dealing with functions in algebraic, graphic, and table form.

Question 1: Recognizing Linear Functions

Which of the following is NOT a linear function?

(A) $f(x) = \dfrac{x}{3} - 6$

(B) $f(x) = 7 - 4x$

(C) $f(x) = 6x$

(D) $f(x) = \dfrac{1}{x} - 5$

(E) $f(x) = 8$

A linear function fits the form $f(x) = mx + b$, where m is the slope and b is the y-intercept. In order to identify the function that is not linear, then, we can find out which function cannot be made to fit the standard form. Choice (A) might be tempting because the variable is divided by a number. Remember that dividing by 3 is the same as multiplying $\dfrac{1}{3}$. You can say $m = \dfrac{1}{3}$. Choice (B) does not appear in the standard form, but it can be rewritten as $f(x) = -4x + 7$, such that $m = -4$ and $b = 7$. Choice (C) might be tempting as well, since nothing is added to the variable term in the function. You can see that this really is a linear function, though, if you consider that $b = 0$. That function is the same as $f(x) = 6x + 0$.

$f(x) = \dfrac{1}{x} - 5$ is not a linear function, though, because a number is divided by the variable. There is no way to make the expression $\dfrac{1}{x} - 5$ fit the form $mx + b$. **So, (D) is the correct answer choice.**

Choice (E) might also seem like a correct answer; unlike the other linear functions, there is no variable term. In fact, this is a linear function with a slope of 0: a straight, horizontal line. Here, $m = 0$, and the function could be written as $f(x) = 0(x) + 8$.

Question 2: Identifying Slope and Intercepts

What are the intercepts of the function $f(x) = 4x - 8$?

(A) (2, 0) and (0, -8)

(B) (4, 0) and (0, 2)

(C) (-8, 0) and (0, 4)

(D) (4, 0) and (0, -8)

(E) (-8, 0) and (0, 2)

The intercepts of a linear function are the points where its graph crosses each axis. You can read the y-intercept of a linear function in standard form right off the equation. b is the y-intercept of $f(x) = mx + b$. So, the y-intercept of $f(x) = 4x - 8$ is -8. That is the point where the graph of the function intercepts the y-axis, so it has an x-coordinate of 0. Therefore, the y-intercept of the function is the point with coordinates $(0, -8)$.

To find the x-intercept of a linear function, you can find its zero. The x-intercept is a point with a y-coordinate of 0, so its x coordinate is the one that makes $f(x) = 0$. You can find the value of that coordinate, then, by solving $4x - 8 = 0$:

$$
\begin{array}{r}
4x - 8 = 0 \\
\underline{+8 \; +8} \\
4x = 8 \\
\underline{\div 4 \quad \div 4} \\
x = 2
\end{array}
$$

Note that x has the value $-\dfrac{b}{m}$. Recall that you can use that expression to find the x-intercept of a linear function in standard form.

So, the x-intercept has the coordinates $(2, 0)$. **That makes (A) the correct choice.** Choices (B) and (D) use the slope $m = 4$ as a coordinate in the x-intercept. Choice (E) includes the correct values, but uses each of them for the wrong intercept.

Question 3: Identifying Linear Functions from Tables

The numbers in the table below fit which function?

x	f(x)
2	9
3	17
5	33
7	49

(A) $f(x) = 4x + 5$

(B) $f(x) = 4x + 13$

(C) $f(x) = 6x + 3$

(D) $f(x) = 8x - 7$

(E) $f(x) = 8x - 5$

The right function is the one where $f(2) = 9, f(3) = 17, f(5) = 33$, and $f(7) = 49$. The function must have the right value for every given value of x. Finding a single row that matches the values given by the function isn't enough. In this question, then, you need to test each function, seeing whether it maps each of the elements in the domain onto the right element in the range.

Choice (A) gives the correct value for the second row: $f(3) = 4(3) + 5 = 12 + 5 = 17$. As for the first row, $f(2) = 4(2) + 5 = 8 + 5 = 13$. Since that's not the value of the function in the table when $x = 2$, you can rule out this choice. In choice (B), $f(5) = 4(5) + 13 = 20 + 13 = 33$. The values of $f(x)$ don't match those in the table for the other values of x. You would find that the functions in choices (C) and (E) don't fit either.

$f(x) = 8x - 7$, on the other hand, gives us these values:

$f(2) = 8(2) - 7 = 16 - 7 = 9$

$f(3) = 8(3) - 7 = 24 - 7 = 17$

$f(5) = 8(5) - 7 = 40 - 7 = 33$

$f(7) = 8(7) - 7 = 56 - 7 = 49$

Choice (D) is the correct answer.

Question 4: Graphing Linear Functions

Which of the following shows the graph $g(x) = \frac{x}{4} - 1$?

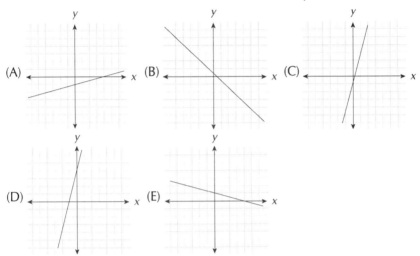

You can identify the graph of a linear function quickly by finding its intercepts. Once you know the intercepts, you can pick out the graph that

has both of them. Recall that for the linear function $g(x) = mx + b$, the x-

intercept is $-\frac{b}{m}$ and the y-intercept is b. In $g(x) = \frac{x}{4} - 1$, $b = -1$ and

$-\frac{b}{m} = -\frac{-1}{\frac{1}{4}} = -(-1 \times 4) = -(-4) = 4$. So, the x-intercept has coordinates

$(4, 0)$, and the y-intercept has coordinates $(0, -1)$. Only the line in choice (A) includes both of those coordinates. **So, choice (A) is the correct answer.**

Choice (C) is the graph of $g(x) = 4x - 1$. Choice (D) is the graph of

$g(x) = \frac{x}{4} + 4$. The x- and y- intercepts are switched there. Choice (E) is the

graph of $g(x) = -\frac{x}{4} + 1$. That graph has a y-intercept of 1 instead of -1.

Question 5: Identifying Linear Functions from Graphs

Which of the following functions is graphed below?

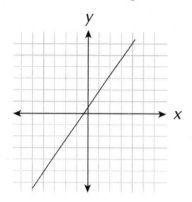

(A) $f(x) = \dfrac{x + 3}{3}$

(B) $f(x) = \dfrac{x + 3}{2}$

(C) $f(x) = \dfrac{2x + 2}{3}$

(D) $f(x) = \dfrac{2x + 1}{3}$

(E) $f(x) = \dfrac{3x + 1}{2}$

To identify a linear function from its graph, you can select two points on the line and use them to find the slope. You can use the slope and the coordinates of one point to find the y-intercept.

The line includes the points $(-1, -1)$ and $(1, 2)$. We can say that $x_1 = -1, x_2 = 1, y_1 = -1$, and $y_2 = 2$, and plug those values into the slope formula:

$$m = \frac{y_2 - y_1}{x_2 - x_1} = \frac{2 - (-1)}{1 - (-1)} = \frac{2 + 1}{1 + 1} = \frac{3}{2}$$

So the slope *m* is $\frac{3}{2}$. Let's take that value and the coordinates of one point, and plug them into $f(x) = mx + b$:

$$x = 1, \ f(x) = 2, \ m = \frac{3}{2}, \text{ so } 2 = \frac{3}{2}(1) + b$$

$b = \frac{1}{2}$ and the function $f(x) = \frac{3x}{2} + \frac{1}{2}$ $\frac{3x}{2} + \frac{1}{2} = \frac{3x+1}{2}$, **choice (E) is the correct answer.** Choice (D) is the result of getting a slope of $\frac{2}{3}$ instead $\frac{3}{2}$.

CHAPTER QUIZ

1. Which of the following is a linear function?

   (A) $f(x) = 3x(4 + 2x)$

   (B) $f(x) = 2x^2 + 7$

   (C) $f(x) = \dfrac{\sqrt{x}}{5} - 7$

   (D) $f(x) = \dfrac{7 - 3x}{5}$

   (E) $f(x) = \dfrac{6}{5x}$

2. Which of the following functions CANNOT be a linear function?

(A)
x	f(x)
2	7
3	10
4	13
5	16

(B)
x	f(x)
2	7
3	11
4	15
5	19

(C)
x	f(x)
2	-3
3	-4
4	-5
5	-6

(D)
x	f(x)
2	5
3	7
4	10
5	13

(E)
x	f(x)
2	8
3	12
4	16
5	20

3. What is the slope of the function $f(x) = \dfrac{4x + 3}{5}$?

   (A) $\dfrac{3}{5}$

   (B) $\dfrac{4}{5}$

   (C) 3

   (D) 4

   (E) 5

4. Which function has a y-intercept of $\frac{1}{3}$?

   (A) $f(x) = 1 - 3x$

   (B) $f(x) = 3 + x$

   (C) $f(x) = \dfrac{x + 5}{3}$

   (D) $f(x) = \dfrac{2 + x}{3}$

   (E) $f(x) = \dfrac{2 + 4x}{3} - \dfrac{1}{3}$

5. What is the x-intercept of $g(x) = 1 + \dfrac{2 + x}{5}$?

   (A) −7

   (B) −2

   (C) 1

   (D) $\dfrac{5}{7}$

   (E) $\dfrac{7}{5}$

6. Which function could the table below represent?

x	f(x)
2	5
4	21
7	45

   (A) $f(x) = 4x + 5$

   (B) $f(x) = 6x - 7$

   (C) $f(x) = 6x + 3$

   (D) $f(x) = 8x - 11$

   (E) $f(x) = 10x - 15$

7. Which table matches the function $f(x) = 7x - 15$?

(A)

x	f(x)
−12	−99
−11	−91
6	27

(B)

x	f(x)
−10	−75
1	−8
9	48

(C)

x	f(x)
−2	−29
4	13
8	31

(D)

x	f(x)
−5	−50
−1	−24
3	6

(E)

x	f(x)
−3	−36
2	−1
7	34

8. Which of the following is the graph of $g(x) = \dfrac{3x}{2} + 2$?

(A) 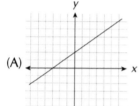 (B) 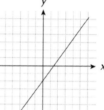 (C)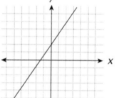

(D) 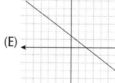 (E)

9. Which function is graphed below?

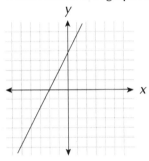

(A) $f(x) = 2x - 2$

(B) $f(x) = 2x + 4$

(C) $f(x) = 4x - 2$

(D) $f(x) = 4x + 2$

(E) $f(x) = 4x + 4$

10. The line on the grid below is the graph of which function?

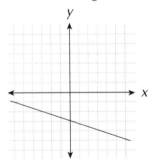

(A) $g(x) = -\dfrac{5x}{2} - 3$

(B) $g(x) = -\dfrac{2x}{5} - 3$

(C) $g(x) = \dfrac{2x}{5} - 3$

(D) $g(x) = \dfrac{2}{5} - 3x$

(E) $g(x) = \dfrac{2}{5} + 3x$

ANSWER EXPLANATIONS

1. D

$f(x) = \dfrac{7 - 3x}{5}$ is the only function that can be made to fit the standard form $f(x) = mx + b$. It can be rewritten as $f(x) = -\dfrac{3x}{5} + \dfrac{7}{5}$. Each of the other functions, if graphed, would appear as curves.

2. D

You can determine whether each function is linear by plotting each set of the ordered pairs on a coordinate grid. If there is no straight line that passes through each point, then the function is not linear. Here is the graph for choice (D):

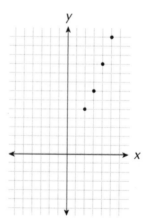

No straight line passes through all four points. Choice (A) fits the linear function $f(x) = 3x + 1$. Choice (B) fits the function $f(x) = 4x - 1$. Choice (C) fits $f(x) = -x - 1$, and (E) fits $f(x) = 4x$.

3. B

In standard form, this function is $f(x) = \dfrac{4x}{5} + \dfrac{3}{5}$. Since m represents the slope in $f(x) = mx + b$, the slope is $\dfrac{4}{5}$. Choice (A), $\dfrac{3}{5}$, is the y-intercept. Choice (D), 4, is the slope of the function $f(x) = 4x + \dfrac{3}{5}$.

4. E

$f(x) = \dfrac{2 + 4x}{3} - \dfrac{1}{3} = \dfrac{2}{3} + \dfrac{4x}{3} - \dfrac{1}{3} = \dfrac{4x}{3} + \dfrac{2}{3} - \dfrac{1}{3} = \dfrac{4x}{3} + \dfrac{1}{3}$. So, that function has a y-intercept of $\dfrac{1}{3}$. Choice (A) has an x–intercept of $\dfrac{1}{3}$. Choices (C) and (D) have slopes of $\dfrac{1}{3}$.

5. A

You could get the x-intercept of $g(x) = 1 + \dfrac{2 + x}{5}$ by putting it in standard form and then finding the value $-\dfrac{b}{m}$. Alternatively, you could find the zero of the function more directly by $0 = 1 + \dfrac{2 + x}{5}$:

$$0 = 1 + \dfrac{2 + x}{5}$$
$$\dfrac{-1 - 1}{-1 = \dfrac{2 + x}{5}}$$

$$\dfrac{\times 5 \quad \times 5}{-5 = 2 + x}$$

$$\dfrac{-2 - 2}{-7 = x}$$

Choice (B) gives $g(x)$ a value of 1 instead of 0. Choice (E) is the y-intercept of the function.

6. D

For $f(x) = 8x - 11$, $f(2) = 8(2) - 11 = 16 - 11 = 5$, $f(4) = 8(4) - 11 = 32 - 11 = 21$, and $f(7) = 8(7) - 11 = 56 - 11 = 45$, so that function matches the values in the table perfectly. Choice (A) fits only the second row of the table. Choice (B) fits the first row, but not the others. Choice (C) fits only the third row.

7. E

If $f(x) = 7x - 15$, then $f(-11) = -77 - 15 = -92$. So, the table in choice (A) does not match the linear function. Since $f(-10) = -70 - 15 = -85$, choice (B) is incorrect. $f(8)$ equals 41, rather than 31, so (C) is incorrect. Choice (D) is incorrect because $f(-1) = -22$, and not -24. The pairs in choice (E), however, match the function:

$f(-3) = -21 - 15 = -36$
$f(2) = 14 - 15 = -1$
$f(7) = 49 - 15 = 34$

8. C

In this linear function, $m = \dfrac{3}{2}$ and $b = 2$. So, the x-intercept

$-\dfrac{b}{m} = -\dfrac{2}{\frac{3}{2}} = -\left(2 \times \dfrac{2}{3}\right) = -\dfrac{4}{3}$. The right graph, then, has coordinates

$\left(-\dfrac{4}{3}, 0\right)$ and $(0, 2)$.

The only graph with these points is in choice (C). Choice (A) is the graph

$g(x) = \dfrac{2}{3} + 2$. Choice (D) is the graph $g(x) = -\dfrac{3}{2} + 2$ which has an x-intercept of $\dfrac{4}{3}$ instead of $-\dfrac{4}{3}$.

9. B

This line has a y-intercept of 4. It has points with coordinates (–2, 0) and

(0, 4). So, the slope of the line is $m = \dfrac{y_2 - y_1}{x_2 - x_1} = \dfrac{4 - 0}{0 - (-2)} = \dfrac{4}{2} = 2$ and the

function is $f(x) = 2x + 4$. Choices (A) and (C) use the value of the
x-intercept in the function instead of the y-intercept.

10. B

The y-intercept of this line is –3. The line has points with coordinates
(0, –3) and (5, –5). Thus, the slope of the line is

$m = \dfrac{y_2 - y_1}{x_2 - x_1} = \dfrac{-5 - (-3)}{5 - 0} = \dfrac{-5 + 3}{5} = -\dfrac{2}{5}$ and the function is

$g(x) = -\dfrac{2x}{5} - 3$. Choice (A) is the result of using the formula $m = \dfrac{x_2 - x_1}{y_2 - y_1}$

instead of $m = \dfrac{y_2 - y_1}{x_2 - x_1}$.

Quadratic Functions

WHAT ARE QUADRATIC FUNCTIONS?

A "quadratic" is a second-order polynomial. That's a polynomial where the highest order term has a variable raised to the second power (the order of an algebraic term is the power to which the variable is raised).

A *quadratic function* is in the form of $f(x) = ax^2 + bx + c$. You will usually find numbers appearing in place of the letters a, b, and c. Those letters, or the numbers that would go in their place, are called *coefficients*. The letter c is also known as the value of the *constant*.

Many quadratic functions feature trinomials (polynomials with three terms), but many do not. Keep in mind that any b or c can have a value of 0. If $a = 3$, $b = 0$, $c = 2$, for instance, we have the quadratic function $f(x) = 3x^2 + 2$ (since $b = 0$, $bx = 0$). As long as a is not 0, you have a quadratic function.

CONCEPTS TO HELP YOU

Quadratics may be familiar to you from Algebra I. We will expand on that topic here by examining some important properties of quadratic functions, including ranges, zeroes, and graphs.

i. Quadratic Equations

Given the function $f(x) = ax^2 + bx + c$, you may need to solve the equation $ax^2 + bx + c = 0$. The solution of such an equation is the zero of the function, as we explained in Chapter 1. Zeroes are also called *roots*, especially in connection with quadratic functions.

ii. Domain and Range

The domain of a quadratic function is typically the set of all real numbers. Every quadratic function has either a minimum or maximum value in the range, however. The graph of a quadratic function is a *parabola*, a U-shaped curve. Each one is shaped like an umbrella, or an upside-down umbrella. Here are two examples:

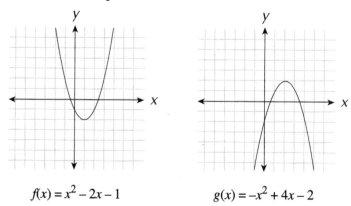

$$f(x) = x^2 - 2x - 1 \qquad\qquad g(x) = -x^2 + 4x - 2$$

The lowest point on $f(x)$ is called the vertex. In $g(x)$, the vertex is the highest point on the curve. The *vertex* is the point on a parabola where the curve changes from decreasing y-values to increasing, or vice versa.

In the graph of $f(x)$, the vertex is (1, –2). Since no y-value is less than –2, the range of the function is $f(x) \geq -2$. In the graph of $g(x)$, on the other hand, the highest point is (2, 2). The range is $g(x) \leq 2$, since no y-value is greater than 2. In the **Steps You Need to Remember** section, we'll go over a technique for finding the highest or lowest point on a parabola when you don't have a graph handy. We'll also examine parabolas further in Chapter 6.

iii. Roots

Many quadratics can be factored into binomials or monomials.

Suppose we have the function $f(x) = ax^2 + bx + c$, and $ax^2 + bx + c$ is the product of $x - j$ and $x - k$. Since the product of two expressions is 0 if one of the expressions has a value of 0, $(x - j) \bullet (x - k) = 0$ if $x - j = 0$ or $x - k = 0$.

Because you get 0 by subtracting any number from itself, $x - j = 0$ if $x = j$, and $x - k = 0$ if $x = k$. So, when you factor the polynomial in $ax^2 + bx + c$ into $x - j$ and $x - k$, j and k are the roots of the quadratic function. Therefore, many equations of the form $ax^2 + bx + c = 0$ have two solutions, and quadratic functions have two roots.

The roots of a quadratic function are the x-intercepts of its graph. The roots of the function are its zeroes, after all, the points on the graph where $y = 0$. Therefore, if you are asked to find the x-intercepts of a quadratic function, you are really being asked to find its roots.

BINOMIAL SQUARES

While many quadratic functions have two roots, some have just one. Since a quadratic function has roots j and k, the function has one root if $j = k$. So, a quadratic function with one root could be rewritten as $f(x) = (x - j) \bullet (x - j)$, or $f(x) = (x - j)^2$. Such quadratic functions involve *binomial squares*.

Here are some binomial squares:

$(x - 4)^2 = x^2 - 8x + 16$

$(x + 7)^2 = x^2 + 14x + 49$

There is one pattern we can recognize here that will help us to identify quadratic functions with binomial squares: you have a binomial square if $\left(\dfrac{b}{2}\right)^2 = c$.

iv. The Quadratic Formula

Many quadratic functions are just too difficult to deal with by factoring. This is often the case when the functions have roots that can be expressed only as radicals. In any case, you can always use the *Quadratic Formula* to find roots. This formula will give you the roots of a quadratic function on the basis of the values of the coefficients you plug in.

Where $ax^2 + bx + c = 0$, $x = \dfrac{-b \pm \sqrt{b^2 - 4ac}}{2a}$

Take the function $f(x) = 2x^2 + 3x + 4 = 0$, for instance. There, $a = 2$, $b = 3$, and $c = 4$. Finding the roots of the function is then simply a matter of evaluating $\dfrac{-b \pm \sqrt{b^2 - 4ac}}{2a}$ for those values of a, b, and c.

Keep in mind that if the quadratic has only two terms, the coefficient of the missing term is 0. In $x^2 - 25 = 0$, for instance, $b = 0$. In $x^2 + 3x = 0$, $c = 0$.

You can evaluate $\dfrac{-b \pm \sqrt{b^2 - 4ac}}{2a}$ to get two roots if your function has that many. That's because the $\pm$ sign in the formula stands for both addition and subtraction. The expression $3 \pm \sqrt{5}$, for example, means $3 + \sqrt{5}$ and $3 - \sqrt{5}$. So, an expression with the $\pm$ sign is actually two expressions rolled into one.

v. Imaginary Numbers and Complex Roots

As you may recall, the radical of a negative, such as $\sqrt{-2}$, has no real number value. There's no number you can square to get -2, after all.

However, the value of the radical of a negative number can be represented with unreal, or *imaginary* numbers. The imaginary unit i is the value of $\sqrt{-1}$. As an imaginary number, it cannot be represented on a graph or number line.

If n is a positive real number, then $\sqrt{-n^2} = \sqrt{(n^2)(-1)} = n\sqrt{-1} = ni$, an imaginary number.

When you use the quadratic formula and get a negative value for $b^2 - 4ac$, you have a negative radical. In that case, $\dfrac{-b \pm \sqrt{b^2 - 4ac}}{2a}$ has no real number value. The quadratic function is likely to have *complex roots* in that case. A *complex number* is the sum of a real number and an imaginary number. If $b \neq 0$, then the roots are complex. If $b = 0$, then the roots are just imaginary. Either way, they are not real.

A quadratic function with no real number roots has no x-intercepts. The vertex is above the axis if the parabola opens up, or it is below the axis if the parabola opens down.

STEPS YOU NEED TO REMEMBER

The following steps are important parts of different methods used to find the roots of quadratic functions. Additionally, we'll explain how to find the vertex and the range of a quadratic function.

i. Graphing Quadratic Functions

Graphing quadratic functions is not as simple as graphing linear ones; you can't just draw a line through two points already plotted. Graphs of quadratic functions are curved. You may have to plot many points to get an accurate parabola. You can plot points by picking values for x, and evaluating the function for each. Once you're able to find the roots and vertex of the graph using the steps that follow, you can make sure that important points on the graph are plotted.

If you plot the intercepts and vertex of the parabola, graphing the rest can be a little more straightforward. That's because parabolas are symmetrical; one half of the curve is a *reflection* of the other. The halves are reflections over a vertical line that intercepts the vertex, called the *axis of symmetry*. We'll look at reflections in more detail in Chapter 9.

ii. Finding the Roots by Graphing

As we've discussed, the roots of a quadratic function (with real number roots) are the x-coordinates of the x-intercepts. If you are having trouble factoring or applying the quadratic formula, you can always try to graph the function. If you can determine exactly where the parabola crosses the axis, you'll know the roots.

iii. Finding the Range

You can find the range of a quadratic function $f(x) = ax^2 + bx + c$ by rewriting it in a different form. It happens that a quadratic can be expressed as the sum of the square of a binomial $x - h$ and a constant k. So,

$$ax^2 + bx + c = (x - h)^2 + k$$

For instance, $x^2 - 8x + 20 = (x^2 - 8x + 16) + 4 = (x - 4)^2 + 4$

The variables h and k are the coordinates of the vertex of the parabola. So, the vertex of $f(x) = x^2 - 8x + 20$ is $(4, 4)$.

To get $(x - h)^2 + k$ from $x^2 + bx + c$, follow these steps:

- Divide b by 2.
- Square $x + \dfrac{b}{2}$.
- Subtract $\left(x + \dfrac{b}{2}\right)^2$ from $x^2 + bx + c$. The difference is k.

$h = -\dfrac{b}{2}$

It is really the value of k that you need for the range, because k is the value of the y-coordinate. Once you have that, you're halfway there. You then need to determine whether the range includes all of the values greater than k, or less than k. That depends on which way the parabola opens, up or down. If the value of a in $f(x) = ax^2 + bx + c$ is positive, then the parabola opens up, and the range is $f(x) \geq k$. If a is negative, then the parabola opens down, and the range is $f(x) \leq k$.

iv. The Quadratic Formula

Remember to get your function into the standard form of $f(x) = ax^2 + bx + c$ before using the quadratic formula. If the terms are out of order, you run the risk of mixing up the values of a, b, and c. Thus, a must be the coefficient of the second-order term, b must be the coefficient of the first-order term, and c must be the value of the constant. Once you have the values in order, you need to evaluate $\dfrac{-b \pm \sqrt{b^2 - 4ac}}{2a}$ and simplify as much as possible.

v. The Discriminant

When you need to know whether a quadratic function has real number roots, there is a shortcut you can apply. If $b^2 - 4ac$ (the expression inside the radical sign in the quadratic formula) is negative, then the roots are complex or imaginary. You can cut to the chase, then, by evaluating $b^2 - 4ac$. That part of the formula is known as the *discriminant*. In general, a quadratic function has real roots only if $b^2 \geq 4ac$.

vi. Factoring

When $a = 1$ in a quadratic function, you can try to factor the quadratic into two binomials, $x - j$ and $x - k$. If you are successful, then you know that j and k are the roots. To find the values of j and k, note that $x^2 + bx + c = x^2 + -(j + k)x + jk$. So $b = -(j + k)$, or $-b = j + k$. Also, $c = jk$. In order to identify the values of j and k, then, you must find the two numbers that have a sum of $-b$ and a product of c.

To deal with functions where a has a value other than 1, let's introduce two more coefficients, d and e. They'll go in front of the variables of the binomial factors. In general,

$(dx - j) \bullet (ex - k) = ax^2 + bx + c$
So, $de = a$, $-dk - ej = b$, and $jk = c$.

You may encounter quadratics with terms that have large coefficients. When that happens, getting the correct values of d, e, j, and k may be a matter of trial and error. You should start by taking different pairs of numbers that have a product of a, and different pairs of numbers that have a product of dk and ej. Try to find a combination that satisfies the equation $dk + ej = -b$.

STEP-BY-STEP ILLUSTRATION OF THE FIVE MOST COMMON QUESTION TYPES

Now that we've reviewed the important concepts and techniques related to quadratic functions, it's time to apply them. Our focus in this chapter has been on the roots, graphs, and ranges of quadratic functions. We'll now walk through helpful approaches to answering questions on those topics.

Question 1: Finding the Range of a Quadratic Function

What is the range of $f(x) = x^2 + 12x + 20$?

(A) $f(x) \leq -16$

(B) $f(x) \geq -16$

(C) $f(x) \geq -6$

(D) $f(x) \leq 6$

(E) $f(x) \geq 16$

You can find the range of this function by getting the y-coordinate of the vertex. That involves rewriting $x^2 + 12x + 20$ in the form of $(x - h)^2 + k$. Since $12 \div 2 = 6$, $x^2 + 12x + 20 = (x + 6)^2 + k$. Now let's get the value of k:

$$k = x^2 + 12x + 20 - (x + 6)^2 = x^2 + 12x + 20 - (x^2 + 12x + 36) = -16$$

So, $f(x) = x^2 + 12x + 20 = (x + 6)^2 - 16$.

$h = -6$ and $k = -16$.

The coordinates of the vertex are $(-6, -16)$. Since the value of a in this function is positive, the parabola opens up. So the range includes -16 and every number greater than -16. **Choice (B)** is the correct answer. Choice (A) uses -16 as the maximum value of the range instead of the minimum. Choice (C) uses the value of h instead of k, and (D) uses $-h$. Choice (E) uses $-k$ instead of k.

Question 2: Factoring

What are the roots of $f(x) = x^2 - 19x + 84$?

(A) 6 and 13

(B) 6 and 14

(C) 7 and 12

(D) 8 and 11

(E) 9 and 10

A quadratic function $f(x) = ax^2 + bx + c$ has roots j and k if $ax^2 + bx + c = (x - j) \cdot (x - k)$. So we can find the values of j and k by factoring the quadratic into two binomials.

$(x - j) \cdot (x - k) = x^2 - (j + k)x + jk$. If $x^2 - (j + k)x + jk = x^2 - 19x + 84$, then $j + k = 19$, and $jk = 84$. Since j and k have a positive sum and product, j and k are both positive.

Here is a list of pairs of positive numbers that have a product of 84, along with their sums:

$1, 84 \rightarrow 1 + 84 = 85$

$2, 42 \rightarrow 2 + 42 = 44$

$3, 28 \rightarrow 3 + 31 = 34$

$4, 21 \rightarrow 4 + 16 = 20$

$6, 14 \rightarrow 6 + 14 = 20$

$7, 12 \rightarrow 7 + 12 = 19$

This last pair, 7 and 12, meets our requirements; the numbers have a sum of 19 and a product of 84. So, $j = 7$ and $k = 12$ (we could have given each variable the other value; the order doesn't matter). We can check this by finding the product of $(x - 7)$ and $(x - 19)$:

$(x - 7) \cdot (x - 12) = (x \cdot x) + (x \cdot -7) + (x \cdot -12) + (-7 \cdot -12) =$
$x^2 - 7x - 12x + 84 = x^2 - 19x + 84$

So, Choice (C) is the correct answer. The numbers in choices (A), (D), and (E) have sums of 19, but don't have the right product. The numbers in choice (B) have a product of 84, but don't have the right sum.

Question 3: Graphing

Which of the following functions is graphed below?

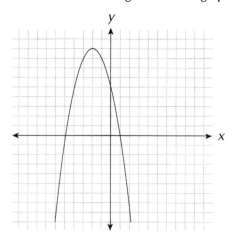

(A) $f(x) = -x^2 - 4x + 5$

(B) $f(x) = -x^2 + 4x + 5$

(C) $f(x) = -x^2 + 6x - 5$

(D) $f(x) = x^2 - 4x + 5$

(E) $f(x) = x^2 + 4x - 5$

You can find the function by using certain coordinates from the graph. Since it crosses the x-axis at $(-5, 0)$ and $(1, 0)$, the roots are -5 and 1. This means that $j = -(-5) = 5$, and $k = -1$. $(x + 5) \cdot (x - 1) = x^2 + 5x - x - 5 = x^2 + 4x - 5$.

We're not done, however. The function $f(x) = x^2 + 4x - 5$ in choice (E), has roots -5 and 1, but it has a graph that opens up. Our graph opens down. In order to get that, you need to multiply the quadratic by -1. The product is $-x^2 - 4x + 5$. **So choice (A) is the correct answer.** Choice (B) has roots 5 and -1, while (C) has roots 1 and 5. Choice (D) has imaginary roots.

Question 4: Using the Quadratic Formula

What are the roots of $f(x) = 2x^2 + 9x - 5$?

(A) $-5, \dfrac{1}{2}$

(B) $\dfrac{-9 - \sqrt{71}}{4}, \dfrac{-9 + \sqrt{71}}{4}$

(C) $\dfrac{-9 - \sqrt{41}}{4}, \dfrac{-9 + \sqrt{41}}{4}$

(D) $\dfrac{-9 - \sqrt{91}}{4}, \dfrac{-9 + \sqrt{91}}{4}$

(E) $5, \dfrac{1}{4}$

Since this quadratic wouldn't be easily factored, you should use the quadratic formula to find the roots. If x is the root of a quadratic function,

then $x = \dfrac{-b \pm \sqrt{b^2 - 4ac}}{2a}$

Since $f(x) = 2x^2 + 9x - 5$, $a = 2$, $b = 9$, and $c = -5$. So,

$$x = \frac{-b \pm \sqrt{b^2 - 4ac}}{2a} = \frac{-9 \pm \sqrt{9^2 - 4(2)(-5)}}{2(2)} = \frac{-9 \pm \sqrt{81 + 40}}{4}$$

$$= \frac{-9 \pm \sqrt{121}}{4} = \frac{-9 \pm 11}{4}$$

$$\frac{-9 + 11}{4} = \frac{2}{4} = \frac{1}{2}$$

$$\frac{-9 - 11}{4} = \frac{-20}{4} = -5$$

So, the roots of the quadratic function are $\frac{1}{2}$ and -5, and **choice (A) is the correct answer.** Choice (B) is the result of using the wrong formula:

$$x = \frac{-b \pm \sqrt{b^2 + 4ac}}{2a}$$

instead of $x = \frac{-b \pm \sqrt{b^2 - 4ac}}{2a}$. Choice (C) is the result of using

$$x = \frac{-b \pm \sqrt{b^2 - ac}}{2a}.$$

Question 5: Finding the Discriminant

Which quadratic function has imaginary roots?

(A) $f(x) = x^2 + 4x + 3$

(B) $f(x) = x^2 + 4x + 5$

(C) $f(x) = 2x^2 + 6x + 4$

(D) $f(x) = 2x^2 + 8x + 6$

(E) $f(x) = 3x^2 + 8x + 4$

A quadratic function $f(x) = ax^2 + bx + c$ has imaginary roots if the discriminant $b^2 - 4ac$ is negative.

So, let's find the discriminant for each function:

$f(x) = x^2 + 4x + 3$:

$$a = 1, b = 4, c = 3$$
$$b^2 - 4ac = (4)^2 - 4(1)(3) = 16 - 12 = 4$$

$f(x) = x^2 + 4x + 5$:

$$a = 1, b = 4, c = 5$$
$$b^2 - 4ac = (4)^2 - 4(1)(5) = 16 - 20 = -4$$

$f(x) = 2x^2 + 6x + 4$:

$\quad$ $a = 2, b = 6, c = 4$

$\quad$ $b^2 - 4ac = (6)^2 - 4(2)(4) = 36 - 32 = 4$

$f(x) = 2x^2 + 8x + 6$:

$\quad$ $a = 2, b = 8, c = 6$

$\quad$ $b^2 - 4ac = (8)^2 - 4(2)(6) = 64 - 48 = 16$

$f(x) = 3x^2 + 8x + 4$:

$\quad$ $a = 3, b = 8, c = 4$

$\quad$ $b^2 - 4ac = (8)^2 - 4(3)(4) = 64 - 48 = 16$

Only $f(x) = x^2 + 4x + 5$ has a negative discriminant. So, that function is the only one with imaginary roots, and **choice (B) is the correct answer.**

CHAPTER QUIZ

1. What is the range of the quadratic function graphed below?

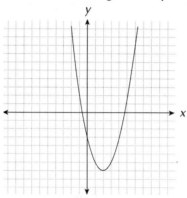

(A) All real numbers
(B) All real numbers less than or equal to –7
(C) All real numbers greater than or equal to –7
(D) All real numbers other than –3
(E) All real numbers other than 2

2. What is the range of $f(x) = x^2 + 8x + 14$?
    (A) $f(x) \leq -4$
    (B) $f(x) \geq -4$
    (C) $f(x) \leq -2$
    (D) $f(x) \geq -2$
    (E) $f(x) \leq 4$

3. What are the zeroes of $f(x) = x^2 + 8x - 105$?
    (A) –15 and 7
    (B) –5 and 21
    (C) –3 and 35
    (D) 5 and 21
    (E) 7 and 15

4. What are the roots of $g(x) = 2x^2 - 12x - 32$?
    (A) −8 and 2
    (B) −4 and 8
    (C) −2 and 8
    (D) 2 and 8
    (E) 4 and 8

5. Which of the following functions is graphed below?

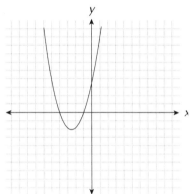

    (A) $f(x) = x^2 - 5x - 4$
    (B) $f(x) = x^2 - 5x + 4$
    (C) $f(x) = x^2 - 4x - 5$
    (D) $f(x) = x^2 + 4x - 5$
    (E) $f(x) = x^2 + 5x + 4$

6. Which of the following is the graph of $f(x) = x^2 + 2x - 8$?

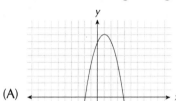

(A)

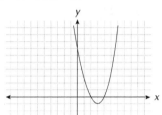

(B)

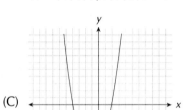

(C)

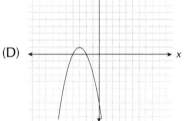

(D)

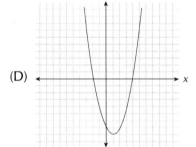

(D)

7. Which of the following is the complete list of the roots of
   $g(x) = x^2 - 8x - 16$?

   (A) −4

   (B) −4, 4

   (C) 4

   (D) $4 \pm 2\sqrt{2}$

   (E) $4 \pm 4\sqrt{2}$

8. What are the x-intercepts of the graph of $f(x) = 8 - 10x - x^2$?

   (A) $-5 \pm \sqrt{33}$

   (B) $-5 \pm 2\sqrt{33}$

   (C) $5 \pm \sqrt{33}$

   (D) $\dfrac{5 \pm \sqrt{33}}{8}$

   (E) $\dfrac{5 \pm 2\sqrt{33}}{8}$

9. Which function has real number roots?

   (A) $f(x) = 3x^2 - 8x + 6$

   (B) $f(x) = 3x^2 + x + 10$

   (C) $f(x) = 4x^2 + 8x + 6$

   (D) $f(x) = 6x^2 - 7x + 2$

   (E) $f(x) = 6x^2 + 6x + 4$

10. Which function has imaginary roots?

    (A) $f(x) = x^2 - 2x - 45$

    (B) $f(x) = x^2 + 9x + 7$

    (C) $f(x) = 2x^2 + 6x + 5$

    (D) $f(x) = 2x^2 - 9x + 10$

    (E) $f(x) = 3x^2 - 6x - 23$

ANSWER EXPLANATIONS

1. C

The vertex of the parabola lies at the point (2, –7). Every other point on the graph is above the vertex. So the range is $f(x) \geq -7$. Choice (A) gives the domain of the function instead of the range. Choice (D) uses the y-intercept of the function, but that value is part of the range. Choice (E) gives the x-coordinate of the vertex, but that number is also part of the range.

2. D

$x^2 + 8x + 14 = (x + 4)^2 + k$. Since $(x + 4)^2 = x^2 + 8x + 16$, $k = -2$. Since a has a positive value, the range is all real numbers greater than or equal to –2. Choice (A) uses the value of h instead of k. Choice (C) uses –2 as the maximum value in the range instead of the minimum.

3. A

The zeroes of a quadratic function are its roots. If $x^2 + 8x - 105 = (x - j) \bullet (x - k)$, then $j + k = -8$ and $jk = -105$. Here is a list of the sums of the number pairs that have a product of 105:

$1 + (-105) = -104$

$3 + (-35) = -32$

$5 + (-21) = -16$

$7 + (-15) = -8$

$15 + (-7) = 8$

$21 + (-5) = 16$

$35 + (-3) = 32$

$105 + (-1) = 104$

Since 7 and –15 have a sum of –8, those are the roots of the function.

4. C

The quadratic $2x^2 - 12x - 32$ can be factored into the binomials $2x + 4$ and $x - 8$. So, $g(x) = 0$ if $2x + 4 = 0$ or $x - 8 = 0$. Since the solutions to those equations are –2 and 8, respectively, those are the zeroes or roots of the function.

Choice (A) gives the roots of $g(x) = 2x^2 + 12x - 32$. Choice (B) gives the roots of $g(x) = 2x^2 - 8x - 64$, while (D) gives the roots of $g(x) = 2x^2 - 20x + 32$. Choice (E) gives the roots of $g(x) = 2x^2 - 24x + 64$.

5. E

The parabola crosses the x-axis at $x = -1$ and $x = -4$. Those numbers are the roots j and k. So the function is $f(x) = (x - j)(x - k) = (x - (-1))(x - (-4)) = (x + 1)(x + 4) = x^2 + x + 4x + 4 = x^2 + 5x + 4$.

Choice (B) is the function with roots 1 and 4, rather than –1 and –4. Choice (D) has roots 1 and –5.

6. C

The quadratic function $f(x) = x^2 + 2x - 8$ has roots –4 and 2, since $x^2 + 2x - 8 = (x + 4) \cdot (x - 2)$. The graph in choice (C) is the only one with those x-intercepts.

Choice (B) is the graph of the quadratic function with roots 2 and 4, instead of 2 and –4. Choice (E) is the graph of $f(x) = x^2 - 2x - 8$.

7. E

You would most likely have difficulty finding the zeroes of this function by factoring. Use the quadratic formula instead. In $g(x) = x^2 - 8x - 16$, $a = 1$, $b = -8$, and $c = -16$. So,

$$x = \frac{-b \pm \sqrt{b^2 - 4ac}}{2a} = \frac{-(-8) \pm \sqrt{(-8)^2 - 4(1)(-16)}}{2(1)} = \frac{8 \pm \sqrt{64 - (-64)}}{2} =$$

$$\frac{8 \pm \sqrt{64 + 64}}{2} = \frac{8 \pm \sqrt{64(2)}}{2} = \frac{8 \pm 8\sqrt{2}}{2} = 4 \pm 4\sqrt{2}$$

Choice (A) is the root of $g(x) = x^2 - 8x + 16$, while choice (B) gives the roots of $g(x) = x^2 - 16$.

8.   A

Remember that the x-intercepts of a quadratic function's graph are the function's roots. So this question is really just asking for the roots. Before you apply the quadratic formula, get the function into a standard form:

$f(x) = 8 - 10x - x^2 = -x^2 - 10x + 8$. So $a = -1$, $b = -10$, and $c = 8$.

$$x = \frac{-b \pm \sqrt{b^2 - 4ac}}{2a} = \frac{-(-10) \pm \sqrt{(-10)^2 - 4(-1)(8)}}{2(-1)} = \frac{10 \pm \sqrt{100 + 32}}{-2} =$$

$$\frac{10 \pm \sqrt{132}}{-2} = \frac{10 \pm \sqrt{4(33)}}{-2} = \frac{10 \pm 2\sqrt{33}}{-2} = -5 \pm \sqrt{33}$$

Choice (D) is the result of not getting the function in standard form first, and switching the values of a and c in the formula.

9.   D

The discriminant $b^2 - 4ac$ of each function is as follows:

(A) $a = 3$, $b = -8$, $c = 6$; $b^2 - 4ac = (-8)^2 - 4(3)(6) = 64 - 72 = -8$
(B) $a = 3$, $b = 1$, $c = 10$; $b^2 - 4ac = (1)^2 - 4(3)(10) = 1 - 120 = -119$
(C) $a = 4$, $b = 8$, $c = 6$; $b^2 - 4ac = (8)^2 - 4(4)(6) = 64 - 96 = -32$
(D) $a = 6$, $b = -7$, $c = 2$; $b^2 - 4ac = (-7)^2 - 4(6)(2) = 49 - 48 = 1$
(E) $a = 6$, $b = 6$, $c = 4$; $b^2 - 4ac = (6)^2 - 4(6)(4) = 36 - 96 = -60$

Only $f(x) = 6x^2 - 7x + 2$ has a positive discriminant. All of the other functions have imaginary roots.

10. C

The function with imaginary roots is the one with a negative discriminant ($b^2 - 4ac < 0$). In choice (C), $a = 2$, $b = 6$, and $c = 5$, such that $b^2 - 4ac = 36 - 40 = -4$. The discriminant is positive in each of the other choices:

(A) $b^2 - 4ac = 4 - (-180) = 184$

(B) $b^2 - 4ac = 81 - 28 = 53$

(C) $b^2 - 4ac = 36 - 40 = -4$

(D) $b^2 - 4ac = 81 - 80 = 1$

(E) $b^2 - 4ac = 36 - (-276) = 312$

Polynomials and Rational Expressions

WHAT ARE POLYNOMIALS AND RATIONAL EXPRESSIONS?

A *polynomial* is an expression consisting of one or more monomials. A *monomial* is a number, a variable, or the product of a number and one or more variables. If a polynomial contains more than one monomial, they are combined with addition or subtraction. A *binomial* is a polynomial with two monomial terms, such as $x + 3$, $x + y$, and $x + x^2$.

A *trinomial* is a polynomial with three monomial terms, such as $3x^2 - 6x - 9$ and $y + yz + xyz$. Polynomials often include powers. Those powers can be used to classify polynomials by their *order*. For instance, $x^3 + 5x^2$ is a third-order polynomial because the largest exponent is 3. Quadratics, as we discussed in the previous chapter, are second-order polynomials.

Note that a third-order polynomial is not always a trinomial. $2x^3 + 4x$ and $4x^3 + x^2 + 3x + 8$ are both third order polynomials, but neither of them have three terms.

A *rational expression* is a fraction with variables, in the numerator, the denominator, or both. In this chapter, we'll focus on simplifying such fractions, and finding the domain of rational expression functions.

CONCEPTS TO HELP YOU

To deal with polynomials in Algebra II, you'll need to carry out addition, subtraction, and multiplication. The most challenging operation, however, is division. We'll take that up along with factoring, which is essential to finding the roots of polynomial functions and simplifying rational expression functions.

i. Addition and Subtraction

You can add and subtract polynomials by combining *like* terms. Two terms are like only if neither contains a variable that the other doesn't have, or a variable raised to a certain power in one but not the other. Furthermore, any variable found in one term must be raised to the same power in both.

For example, $2x^4$ and $5x^4$ are like terms. They can be added or subtracted. $6a^3b$ and $-7a^3b$ are another pair of like terms. y^2 and $2y^3$, on the other hand, are not like terms.

ii. Multiplication

Finding the product of two polynomials is a matter of pairing each term in the first polynomial with each term in the second, and finding the product of each pair.

Take the polynomials $a + b$ and $c + d + e$:

$$(a + b) \bullet (c + d + e) = a \bullet (c + d + e) + b \bullet (c + d + e)$$

To multiply $c + d + e$ by a, you must multiply a by each term in the trinomial:

$$a \bullet (c + d + e) = ac + ad + ae$$

Likewise,

$$b \bullet (c + d + e) = bc + bd + be$$

So the product of $a + b$ and $c + d + e$ is $ac + ad + ae + bc + bd + be$.

iii. Division and Factoring

One polynomial is evenly divisible by another if the second is a factor of the first. Since $3x$ is a factor of $6x^2 + 12x$, you can divide $6x^2 + 12x$ by $3x$. It is just a matter of common factors canceling out:

$\dfrac{6x^2 + 12x}{3x} = \dfrac{3x(2x + 4)}{3x}$. We can divide the top and bottom of the fraction by $3x$ to eliminate this common factor, leaving just $2x + 4$. This shows that $(6x^2 + 12x) \div 3x = 2x + 4$.

There are many cases where polynomials have no monomial factors, however. You may have to use other methods to extract a polynomial's factors, which could have two or more terms. In the previous chapter, we reviewed techniques for finding the factors of a quadratic. In the **Steps You Need to Remember** section, we'll explain some techniques you can use to find the factors of higher order polynomials.

iv. Roots, Zeroes, and Factors

The roots or zeroes of a polynomial function $f(x)$ are the values of x for which $f(x) = 0$. Many function questions in Algebra II involve finding the roots of polynomial functions. They can also involve intercepts, as the roots of a function represent the points where its graph crosses the x-axis.

As with quadratics, polynomial roots are related to factors. A polynomial function has a value of zero whenever one of its factors equals zero. Factoring a polynomial is a common approach to finding the roots of a function. Also, some Algebra II questions will ask you to factor polynomials, independent of any functions.

v. Domains of Rational Expression Functions

There are many rational expression functions whose domains do not include all rational numbers. If the denominator is an algebraic expression that equals zero for some value of the denominator, then that value is not part of the domain. Remember, a function must have a real number value. An argument that would make the function undefined cannot be part of the domain.

vi. Simplifying Rational Expressions

You can often simplify rational expressions by dividing both the numerator and denominator by a common factor. That factor cancels out, leaving the remaining polynomials as parts of a simplified fraction. For example, we can simplify the fraction $\dfrac{6x^2 + 9x}{10x + 15}$ by factoring each binomial, so as to identify a common factor:

$\dfrac{6x^2 + 9x}{10x + 15} = \dfrac{3x(2x + 3)}{5(2x + 3)}$. Since $2x + 3$ is a common factor of $6x^2 + 9$ and

$10x + 15$, we can take it out of both the numerator and the denominator. So

$\dfrac{3x}{5}$ is the simplified form of $\dfrac{6x^2 + 9x}{10x + 15}$.

STEPS YOU NEED TO REMEMBER

As we said earlier, polynomial division can be challenging. We'll explain how to approach it, in connection with factoring. Before that, we'll review the key points of addition, subtraction, and multiplication.

i. Operations

When adding and subtracting polynomials, you can organize monomial terms using the associative and commutative properties in order to combine like terms. For instance,

$(3x + 6y) + (4x + 8y) =$

$(3x + 6y) + 4x + 8y =$

$(3x + 4x) + 6y + 8y =$

$7x + 6y + 8y =$

$7x + 14y$

When dealing with polynomials with many kinds of terms, using the associative and commutative properties to reorganize the polynomials in small steps will help you to make sure that you do not overlook or mismatch any terms.

Remember also that you must distribute the negative sign when subtracting a polynomial. For instance, $x^2 - (2x^2 + 4x)$ equals $x^2 - 2x^2 - 4x$, not $x^2 - 2x^2 + 4x$.

Multiplication, on the other hand, does not involve combining like terms. Rather, every pair of terms gets combined. If you are multiplying one polynomial by another, then every term in the first polynomial gets multiplied by each term in the second. After that, you simplify the results where possible.

ii. Factoring with the Rational Roots Test

To find the factors of a polynomial, or the zeroes of a polynomial function, you can apply the *rational roots test*. Take the lowest order term of the polynomial and determine its factors, positive and negative. Then divide each of those factors by the coefficient of the highest order term.

Take the polynomial $2x^3 - x^2 - 8x + 4$. The factors of the constant term are $-4, -2, -1, 1, 2$, and 4. Dividing those numbers by the coefficient of the highest order term, 2 from $2x^3$, gets you:

$$-2, -1, -\frac{1}{2}, \frac{1}{2}, 1, 2$$

Each of these numbers could be a zero of the function $f(x) = 2x^3 - x^2 - 8x + 4$. Likewise, $x - j$, where j is one of those numbers, is a factor of $2x^3 - x^2 - 8x + 4$. To find out which numbers work, evaluate the polynomial for each. Those that make $f(x) = 0$ are roots of the function, and could be the value of j in the factor $x - j$.

iii. Synthetic Division

Synthetic division works much like long division in arithmetic. You can use it to divide a polynomial by a binomial of the form $x - a$, where a is a constant.

Start by writing out the coefficients of the larger polynomial in a row. If the polynomial is missing a term of a certain order, use 0 as the coefficient. You will use that row with the constant in the binomial.

Suppose you are dividing $2x^3 + 13x^2 + 19x + 20$ by $x + 5$. Put the coefficients of the binomial in a row, and put the constant of the binomial to the left of it:

−5	2	13	19	20

The reason that you put −5 instead of 5 there is that you use the constant a in $x - a$. $x + 5 = x - (-5)$, so that $a = -5$, and $x - a = x + 5$.

Now, bring down the first coefficient, 2.

−5	2	13	19	20
	2			

Multiply the constant −5 by the coefficient you just brought down, and put the product under the next coefficient.

−5	2	13	19	20
		−10		
	2			

Add the numbers in that column:

−5	2	13	19	20
		−10		
	2	**3**		

Now repeat the process for the rest of the columns.

−5	2	13	19	20
		−10	−15	−20
	2	3	4	0

So we have the numbers 2, 3, 4, and 0. You always discard the last number, which will be zero if the polynomials divide evenly. The remaining numbers are the coefficients of the quotient. The resulting polynomial, therefore, is $2x^2 + 3x + 4$.

iv. Finding the Domain of Rational Expression Functions

As we explained in the **Concepts to Help You** section, the denominator of a rational expression function $f(x)$ cannot equal zero. So any value of x that would make the denominator have that value is not part of the domain. You can set up an equation with the denominator on one side and zero on the other side, and solve. The solutions are not part of the domain.

STEP-BY-STEP ILLUSTRATION OF THE FIVE MOST COMMON QUESTION TYPES

Now that we've gone through the essentials of polynomials, we'll apply the key concepts and techniques to five major question types. Not every polynomial we'll deal with here appears in a function, but the steps we'll follow would certainly apply to function questions.

Question 1: Addition, Subtraction, and Multiplication

$((5x^2 + 6) - (3x^2 + 5x)) \bullet (4x^2 + 7x + 8) =$

(A) $8x^3 + 27x^2 + 2x + 48$

(B) $8x^4 - 6x^3 + 5x^2 + 2x + 48$

(C) $8x^3 - 6x^2 + 35x + 48$

(D) $8x^4 - 6x^3 + 47x^2 + 82x + 48$

(E) $8x^4 + 25x^2 + 82x + 48$

This question requires us to both subtract and multiply polynomials. First, we must subtract the binomial $3x^2 + 5x$ from $5x^2 + 6$. The result will be multiplied by $4x^2 + 7x + 8$.

Performing subtraction with the binomials is a matter of combining like terms. Here, the only pair of like terms is the second-order monomials $5x^2$ and $3x^2$.

$$((5x^2 + 6) - (3x^2 + 5x)) = (5x^2 - 3x^2) + 6 - 5x = 2x^2 - 5x + 6$$

Now we multiply $2x^2 - 5x + 6$ by $4x^2 + 7x + 8$. Multiply each term in the first binomial by each term in the second, and combine the products:

$$2x^2 \bullet (4x^2 + 7x + 8) - 5x \bullet (4x^2 + 7x + 8) + 6 \bullet (4x^2 + 7x + 8) =$$

$$(8x^4 + 14x^3 + 16x^2) - (20x^3 + 35x^2 + 40x) + (24x^2 + 42x + 48)$$

Now, subtract the first two polynomials. Combine like terms:

$$(8x^4 + 14x^3 + 16x^2) - (20x^3 + 35x^2 + 40x) = 8x^4 + (14x^3 - 20x^3) +$$
$$(16x^2 - 35x^2) - 40x = 8x^4 - 6x^3 - 19x^2 - 40x$$

Finally, $(8x^4 - 6x^3 - 19x^2 - 40x) + (24x^2 + 42x + 48) =$
$$8x^4 - 6x^3 + (-19x^2 + 24x^2) + (-40x + 42x) + 48 =$$
$$8x^4 - 6x^3 + 5x^2 + 2x + 48$$

So, choice (B) is the correct answer.

Choice (D) is what you might get by not distributing the subtraction sign when subtracting $20x^3 + 35x^2 + 40x$. Remember that you need to subtract all three terms, not just the first one. Choice (E) is the result of multiplying the exponents instead of adding when getting the product of the polynomials.

Question 2: Factoring a Polynomial

Which binomial is a factor of $x^3 + 2x^2 - 21x + 18$?

(A) $x - 9$

(B) $x - 2$

(C) $x + 1$

(D) $x + 3$

(E) $x + 6$

$x - j$ is a factor of a polynomial that equals 0 when $x = j$. The quickest way to answer this question, then, is to test the value of j in each answer choice. That is the rational roots test we explained in the **Steps You Need to Remember** section.

Take $x - 9$. That binomial is a factor if $x^3 + 2x^2 - 21x + 18 = 0$ when $x = 9$. $(9)^3 + 2(9)^2 - 21(9) + 18 = 729 + 162 - 189 + 18 = 720$, so $x - 9$ is not a factor.

Next is $x - 2$. What is the value of the polynomial if $x = 2$? $(2)^3 + 2(2)^2 - 21(2) + 18 = 8 + 8 - 42 + 18 = -8$, so $x - 2$ is not a factor.

To see whether $x + 1$ is a factor, evaluate $x^3 + 2x^2 - 21x + 18$ where $x = -1$: $(-1)^3 + 2(-1)^2 - 21(1) + 18 = -1 + 2 - 21 + 18 = -2$, so we can rule out that binomial.

Next, we evaluate the polynomial when $x = -3$: $(-3)^3 + 2(-3)^2 - 21(3) + 18 = -27 + 18 - 63 + 18 = -54$, so $x + 3$ is not a factor.

Finally, we have $x + 6$. If $x = -6$, then: $x^3 + 2x^2 - 21x + 18 = (-6)^3 + 2(-6)^2 - 21(-6) + 18 = -216 + 72 + 126 + 18 = 0$. So, $x + 6$ is a factor, and **choice (E) is the correct answer.**

Remember that if the factor is $x - j$, you must test j, not $-j$. If you tested $-j$, you might have gotten choices (C) or (D), since $x - 1$ and $x - 3$ also happen to be factors of $x^3 + 2x^2 - 21x + 18$.

Question 3: Finding the Intercepts of the Graph of a Polynomial

What are the *x*-intercepts of $f(x) = 2x^3 + 5x^2 - 18x - 45$?

(A) $-3, -\frac{5}{2}, 3$

(B) $1, -\frac{5}{2}, 9$

(C) $-1, \frac{5}{2}, 9$

(D) $-3, -5, 3$

(E) $-3, 5, 3$

The *x*-intercepts of the graph of a polynomial are its roots or zeroes. Thus, the correct choice is the one where each number listed is a root. The answer choices in this question include many numbers; there are nine numbers to test altogether. Instead of applying the rational roots test to all of them, you can be methodical. Once you've found one, you can rule out the answer choices that don't include it, and focus on the numbers in the remaining choices.

First, take –3. If $f(-3) = 2(-3)^3 + 5(-3)^2 - 18(-3) - 45 = -54 + 45 + 54 - 45 = 0$. So –3 is a zero of $f(x)$. Since choices (B) and (C) don't include that number, they can be eliminated. Each of the remaining choices, (A), (D), and (E), include 3, so there is no need to test that one. Try $-\frac{5}{2}$ next:

$$f\left(-\frac{5}{2}\right) = 2\left(-\frac{5}{2}\right)^3 + 5\left(-\frac{5}{2}\right)^2 - 18\left(-\frac{5}{2}\right) - 45 =$$
$$2\left(-\frac{125}{8}\right) + 5\left(\frac{25}{4}\right) + 45 - 45 =$$
$$-\frac{125}{4} + \frac{125}{4} = 0$$

Since $f\left(-\frac{5}{2}\right) = 0$, $-\frac{5}{2}$ is also a root. Choices (D) and (E) must be incorrect, then, because they don't include that number. **Choice (A) is the correct answer.**

Question 4: Finding the Domain of a Rational Expression

What is the domain of $f(x) = \dfrac{2x^2 + 8x + 6}{2x^2 - 10x - 28}$?

(A) All real numbers other than –7 and 4

(B) All real numbers other than –4 and 7

(C) All real numbers other than –2 and –3

(D) All real numbers other than –2 and 7

(E) All real numbers other than 1 and 3

The function does not have a real number value if the denominator equals 0. A fraction with such a denominator is undefined. Therefore, the domain does not include any value that would make $2x^2 - 10x - 28$ equal to 0. To find those values, solve the quadratic equation $2x^2 - 10x - 28 = 0$.

The coefficient of the first term is 2, but we can simplify things a little by factoring 2 out of the quadratic. Because $2x^2 - 10x - 28 = 2(x^2 - 5x - 14)$, we can divide both sides of the equation by 2 to get

$$x^2 - 5x - 14 = 0$$

Since the numbers –7 and 2 have a sum of –5 and a product of –14,

$$x^2 - 5x - 14 = (x - 7)(x + 2)$$

So $x^2 - 5x - 14 = 0$ if $x - 7 = 0$ or $x + 2 = 0$.

The solutions of the quadratic equation are $x = 7$ and $x = -2$. The domain does not include those numbers, and **choice (D) is the correct answer.**

Choice (E) represents the values for which $2x^2 + 8x + 6 = 0$. If x equals 1 or 3, then $f(x) = 0$. That's not the same as being undefined, though. So those numbers are part of the domain.

Question 5: Simplifying a Rational Expression

$$\frac{x^2 + x - 12}{x^3 + x^2 - 22x - 40} =$$

(A) $\dfrac{x - 5}{x^2 + 2x - 20}$

(B) $\dfrac{x - 3}{x^2 - 3x - 10}$

(C) $\dfrac{x - 2}{x^2 - x - 20}$

(D) $\dfrac{x + 2}{x^2 - x - 20}$

(E) $\dfrac{x + 4}{x^2 - 3x - 10}$

If this algebraic fraction can be simplified, then the numerator and denominator must share a least one common factor. Since Because the numerator is a quadratic, we can find its factors quickly. They are $x - 3$ and $x + 4$. Now since -3 is not a factor of -40, the constant term in the denominator, $x - 3$ is probably not a factor of that polynomial. On the other hand, 4 is a factor of -40, so $x + 4$ is more likely to be a factor. Let's use synthetic division to find out.

We put -4 in the upper-left box, since we have to divide the polynomial by $x - j$. The next four numbers are the coefficients of $x^3 + x^2 - 22x - 40$.

−4	1	1	−22	−40	

Next, we bring down the first coefficient, and multiply it by –4. We put the product under the next coefficient.

–4	1	1	–22	–40	
	↓	–4			
	1				

We then add the number in the next column, putting the sum underneath. We multiply the sum by –4, and repeat the process until we've gone through the numbers.

–4	1	1	–22	–40	
		–4	12	40	
	1	–3	–10	0	

The numbers below the line are the coefficients of the quotient. We discard the last zero, and use the 1, –3, and –10 to get $x^2 - 3x - 10$. We have shown that $x^3 + x^2 - 22x - 40$ is evenly divisible by $x + 4$. That binomial is a common factor of the numerator and the denominator.

Since $x^2 - 3x - 10$ can be factored into $x - 5$ and $x + 2$, $x - 3$ is not another factor of the denominator. Therefore,

$$\frac{x^2 + x - 12}{x^3 + x^2 - 22x - 40} = \frac{(x - 3)(x + 4)}{(x^2 - 3x - 10)(x + 4)} = \frac{x - 3}{x^2 - 3x - 10}$$

Choice (B) is the correct answer.

Choice (A) includes another factor of the denominator, $x - 5$, and factors that binomial out of $x^3 + x^2 - 22x - 40$. Choice (D) does the same thing with $x + 2$, and (E) with $x + 4$.

CHAPTER QUIZ

1. $(x^2 + 5) \cdot (2x - 6) + (3x^3 + 2x + 4) =$

(A) $2x^3 - 3x^2 + 12x - 26$
(B) $2x^3 - 6x^2 + 7x - 26$
(C) $5x^3 - 6x^2 + 12x - 26$
(D) $5x^3 - 6x^2 + 7x - 34$
(E) $5x^3 - 4x^2 + 10x - 34$

2. $(4x^3 + 8) + (4x + 7) \cdot (2x^2 + x - 9) =$

(A) $8x^3 + 14x^2 - 25x - 91$
(B) $8x^3 + 18x^2 - 29x - 55$
(C) $12x^3 + 14x^2 - 25x - 55$
(D) $12x^3 + 18x^2 - 29x - 55$
(E) $12x^3 + 18x^2 + 7x - 91$

3. $x - 6$, $x + 3$, and $x + 5$ are all factors of which polynomial?

(A) $x^3 - 4x^2 - 33x - 70$
(B) $x^3 - 4x^2 + 3x - 90$
(C) $2x^3 - 4x^2 - 6x - 180$
(D) $2x^3 + 4x^2 - 66x - 180$
(E) $2x^3 + 8x^2 - 6x - 180$

4. Which of the following is a factor of $x^3 - 3x^2 - 70x + 144$?

(A) $x - 8$
(B) $x - 2$
(C) $x + 6$
(D) $x + 9$
(E) $x + 12$

5. Which of the following is a y-intercept of $f(x) = x^3 - 8x^2 + 6x - 48$?

(A) -48
(B) -12
(C) -6
(D) 8
(E) 24

6. Which of the following is an x-intercept of the graph of $f(x) = x^3 + 64$?

(A) -8
(B) -4
(C) 2
(D) 16
(E) 64

7. What is the domain of
$$f(x) = \frac{3x^2 - 2x - 5}{4x^2 - 8x - 21}?$$

(A) All real numbers other than $-\frac{7}{4}$ and 3

(B) All real numbers other than $-\frac{5}{3}$ and 1

(C) All real numbers other than $-\frac{3}{2}$ and $\frac{7}{2}$

(D) All real numbers other than 3 and $\frac{7}{2}$

(E) All real numbers other than $\frac{7}{4}$ and $\frac{3}{7}$

8. What is the domain of
$$g(x) = \frac{x^2 + 15x + 54}{x^3 - 13x^2 + 52x - 60}?$$

(A) All real numbers other than -9 and -6

(B) All real numbers other than $-2, -5$, and 6

(C) All real numbers other than 2, 5, and -6

(D) All real numbers other than 2, 5, and 6

(E) All real numbers other than 6 and 9

9. Which of the following is a simplified form of
$$\frac{4x^4 - 8x^3 + 5x - 10}{x - 2}?$$

(A) $4x^3 - 5$

(B) $4x^3 + 5$

(C) $4x^3 + 2x + 5$

(D) $4x^3 + 2x^2 + 5$

(E) $4x^3 + 4x^2 + 5$

10. $\dfrac{2x^2 + 15x + 7}{2x^3 - x^2 - 113x - 56} =$

(A) $\dfrac{1}{x - 8}$

(B) $\dfrac{1}{x - 7}$

(C) $\dfrac{1}{x + 7}$

(D) $\dfrac{1}{2x - 1}$

(E) $\dfrac{1}{2x + 1}$

ANSWER EXPLANATIONS

1. C

First, multiply the binomials:

$$(x^2 + 5) \bullet (2x - 6) = 2x^3 - 6x^2 + 10x - 30$$

Next, find the sum of this product and the other polynomial:

$$(2x^3 - 6x^2 + 10x - 30) + (3x^3 + 2x + 4) = 5x^3 - 6x^2 + 12x - 26$$

Choice (A) is the result of adding $3x^2$ instead of $3x^3$. Choice (B) involves that error too, and also that of getting $5x$ instead of $10x$ when multiplying 5 by $2x$ in the first step. Choices (D) and (E) get -34 instead of -26 upon adding 4 to -30. Choice (E) also involves adding $2x^2$ instead of $2x$.

2. D

The order of operations rule calls for carrying out multiplication first:

$$(4x + 7) \bullet (2x^2 + x - 9) = 8x^3 + 4x^2 - 36x + 14x^2 + 7x - 63 = 8x^3 + 18x^2 - 29x - 63$$

Add the sum to the other binomial:

$$(4x^3 + 8) + 8x^3 + 18x^2 - 29x - 63 = 12x^3 + 18x^2 - 29x - 55$$

Choice (C) is the result of getting $4x$ instead of $4x^2$ when multiplying. Choice (E) is the result of getting -36 instead of $-36x$ when multiplying.

3. D

If the three binomials are all factors of a certain polynomial, then their product must be a factor of the polynomial, or be the polynomial itself. So you should find the product of the binomials:

$$(x - 6) \bullet (x + 3) \bullet (x + 5) = (x^2 - 3x - 18) \bullet (x + 5) = x^3 + 2x^2 - 33x - 90$$

This product is not found in the answer choices. Notice that some of the polynomials begin with a coefficient of 2. If you multiply $x^3 + 2x^2 - 33x - 90$ by 2, the result is $2x^3 + 4x^2 - 66x - 180$.

Choice (C) is the result of getting $15x$ instead of $-15x$ when multiplying by $x + 5$. Choice (E) is the result of getting $x^2 + 3x - 18$ instead of $x^2 - 3x - 18$ when multiplying $x - 6$ and $x + 3$.

4. B

To find the binomial factor $x - j$ of $x^3 - 3x^2 - 70x + 144$, evaluate the polynomial for each value of x such that $x - j = 0$:

$x = 8 \rightarrow (8)^3 - 3(8)^2 - 70(8) + 144 = 512 - 192 - 560 + 144 = -96$

$x = 2 \rightarrow (2)^3 - 3(2)^2 - 70(2) + 144 = 8 - 12 - 140 + 144 = 0$

$x = -6 \rightarrow (-6)^3 - 3(-6)^2 - 70(-6) + 144 = -216 - 108 + 420 + 144 = 204$

$x = -9 \rightarrow (-9)^3 - 3(-9)^2 - 70(-9) + 144 = -729 - 243 + 630 + 144 = -198$

$x = -12 \rightarrow (-12)^3 - 3(-12)^2 - 70(-12) + 144 = -1{,}728 - 432 + 840 + 144 = -1{,}176$

Since $x^3 - 3x^2 - 70x + 144 = 0$ if $x = 2$, $x - 2$ is a factor of the polynomial. Choices (A) and (D) might have been tempting, since the polynomial equals 0 if $x = -8$ or $x = 9$. Those represent the polynomials $x + 8$ and $x - 9$, rather than $x - 8$ and $x + 9$.

5. A

The y-intercept of a polynomial function $f(x)$ is the value where $x = 0$. Here, $f(0) = (0)^3 - 8(0)^2 + 6(0) - 48 = -48$. Choice (C), –6, is one of the roots; that is, an x-intercept of the function.

6. B

An x-intercept of $f(x)$ has a value for which $f(x) = 0$. If $x^3 + 64 = 0$, then $x^3 = -64$. Since -4 is the cube root of -64, -4 is the solution of that equation, and an x-intercept of $f(x)$.

Choice (A), -8, is an x-intercept of $f(x) = x^2 - 64$. Choice (E) 64, is the y-intercept of the function.

7. C

The domain of $f(x)$ cannot include any value for which the denominator of $\dfrac{3x^2 - 2x - 5}{4x^2 - 8x - 21}$ equals 0. The quadratic $4x^2 - 8x - 21$ is the product of $2x + 3$ and $2x - 7$. Since $4x^2 - 8x - 21 = 0$ if $2x + 3 = 0$ or $2x - 7 = 0$, x cannot have the values of $-\dfrac{3}{2}$ or $\dfrac{7}{2}$.

Choice (A) gives the solutions of $4x^2 - 5x - 21 = 0$, rather than $4x^2 - 8x - 21 = 0$. Choice (B) gives the values of x for which the numerator of the function, rather than the denominator, equals 0.

8. D

The domain of $g(x)$ does not include any value of x for which the denominator equals 0. So the numbers the domain does not include are the solutions of $x^3 - 13x^2 + 52x - 60 = 0$. The possible roots are the factors of the constant, -60: $-60, -30, -20, -15, -12, -10, -6, -5, -4, -3, -2, -1, 1, 2, 3, 4, 5, 6, 10, 12, 15, 20, 30,$ and 60. Since neither -9 nor 9 are on that list, rule out choices (A) and (E). Those are actually the values of x for which the numerator equals 0.

Evaluate the denominator for the other values in the answer choices:

-6: $x^3 - 13x^2 + 52x - 60 = (-6)^3 - 13(-6)^2 + 52(-6) - 60 = -216 - 468 - 312 - 60 = -1{,}056$

-5: $x^3 - 13x^2 - 52x - 60 = (-5)^3 - 13(-5)^2 + 52(-5) - 60 = -125 - 325 - 260 - 60 = -770$

-2: $x^3 - 13x^2 - 52x - 60 = (-2)^3 - 13(-2)^2 + 52(-2) - 60 = -8 - 52 - 104 - 60 = -244$

2: $x^3 - 13x^2 - 52x - 60 = (2)^3 - 13(2)^2 + 52(2) - 60 = 8 - 52 + 104 - 60 = 0$

5: $x^3 - 13x^2 - 52x - 60 = (5)^3 - 13(5)^2 + 52(5) - 60 = 125 - 325 + 260 - 60 = 0$

6: $x^3 - 13x^2 - 52x - 60 = (6)^3 - 13(6)^2 + 52(6) - 60 = 216 - 468 + 312 - 60 = 0$

So $x^3 - 13x^2 + 52x - 60 = 0$ when x equals 2, 5, or 6. Those numbers are not part of the domain of $g(x)$.

9. B

Simplify this algebraic fraction by dividing the numerator by the denominator. Use synthetic division:

2	4	−8	0	5	−10
		8	0	0	10
	4	0	0	5	0

So the quotient is $4x^3 + 5$.

10. A

The factors of $2x^2 + 15x + 7$ are $2x + 1$ and $x + 7$. You can try to divide the denominator by each of those binomials, in order to simplify the fraction. Start with $x + 7$, and use synthetic division:

−7	2	−1	−113	−56
		−14	105	56
	2	−15	−8	0

So $\dfrac{2x^2 + 15x + 7}{2x^3 - x^2 - 113x - 56} = \dfrac{(2x + 1)(x + 7)}{(2x^2 - 15x - 8)(x + 7)} = \dfrac{2x + 1}{2x^2 - 15x - 8}$

$2x^2 - 15x - 8$ factors into $2x + 1$ and $x - 8$. So,

$\dfrac{2x + 1}{2x^2 - 15x - 8} = \dfrac{2x + 1}{(2x + 1)(x - 8)} = \dfrac{1}{x - 8}$

Choice (E) leaves $2x + 1$ instead of $x - 8$ as the factor in the denominator.

CHAPTER 5

Matrices

WHAT ARE MATRICES?

A *matrix* is a set of numbers, organized into rows and columns. Values organized by rows and columns are known as *arrays*, and a matrix is a rectangular array. It is rectangular in the sense that all rows are of equal length, as are all columns. Each value in a matrix is called an *element*. Matrices (the plural of matrix) usually appear with square brackets. All of the following are matrices:

$$A = \begin{bmatrix} 2 & 3 \\ 6 & 4 \\ 8 & 9 \end{bmatrix} \qquad B = \begin{bmatrix} -1 & 2 \\ -7 & 5 \end{bmatrix} \qquad C = \begin{bmatrix} 5 & 3 & 20 & 7 \end{bmatrix}$$

Matrices are usually identified with uppercase letters. They can also be described in terms of their dimensions. Since A has three rows and two columns, it is a 3 x 2 (three by two) matrix. B is a 2 x 2 matrix, and C is a 1 x 4 matrix.

A matrix with the same number of rows and columns is called a square matrix. Since B is a 2 x 2 matrix, it is a square matrix.

CONCEPTS TO HELP YOU

Matrix questions in Algebra II can involve standard operations with matrices, such as addition and multiplication. Other kinds of operations are more unique to matrices. They involve performing operations on matrix rows. You can do that to find the inverses of matrices, and to use matrices to solve systems of equations.

i. Adding and Subtracting Matrices

Matrix addition and subtraction are very straightforward operations. Each operation is simply a matter of combining the corresponding elements in the two matrices. If you are finding the sum of two 2 x 2 matrices, you add the top left elements in each matrix, the bottom right elements in each, and so on. The result is a matrix of the same size as the two with which you began.

There is one important restriction on adding or subtracting matrices: the two matrices you combine must be the same size. You can add two 3 x 3 matrices, or two 4 x 2 matrices. You cannot add a 3 x 3 matrix to a 2 x 2 matrix, or a 3 x 4 matrix to a 4 x 3 matrix.

ii. Multiplying Matrices

There are two kinds of multiplication involving matrices: *scalar multiplication* and *matrix multiplication*. The first kind is very straightforward: you multiply each element in the matrix by a given number. The second kind is much more complicated: you actually multiply one matrix by another. This requires you to combine the rows of one matrix with the columns of the other. Organizing the results can be tricky. We'll go through the process in the **Steps You Need to Remember** section.

One important difference between matrix multiplication and matrix addition is that two matrices can be multiplied only if the first has the same number of rows as the second has columns.

iii. Identity Matrices

An identity matrix is a square matrix where the elements along the "diagonal" have a value of 1, and all of the other elements have a value of 0. Each of these is an identity matrix:

$$\begin{bmatrix} 1 & 0 \\ 0 & 1 \end{bmatrix} \quad \begin{bmatrix} 1 & 0 & 0 \\ 0 & 1 & 0 \\ 0 & 0 & 1 \end{bmatrix} \quad \begin{bmatrix} 1 & 0 & 0 & 0 \\ 0 & 1 & 0 & 0 \\ 0 & 0 & 1 & 0 \\ 0 & 0 & 0 & 1 \end{bmatrix}$$

A nonsquare matrix does not have a diagonal that begins in the top row and first column and ends in the bottom row and last column. Therefore, an identity matrix cannot be anything but a square.

iv. Inverse Matrices

Square matrices have multiplicative inverses. While the product of a real number and its inverse is one, the product of a square matrix and its inverse is an identity matrix. Getting the inverse of a number is just a matter of converting it to a fraction, and swapping the numerator and denominator. Getting the inverse of a matrix, on the other hand, is a much more complicated process. We'll cover it in the **Steps You Need to Remember** section.

v. Row Operations

Row operations come in handy when it comes to finding the inverse of a matrix, as you'll see in the **Steps You Need to Remember** section. You can alter matrices by performing operations on their rows. You can interchange rows, multiply rows, add the values in one row to those in another, and so on.

vi. Augmented Matrices

When you use row operations to find the inverse of a matrix, you'll use what is called an *augmented matrix*. Such a matrix is one where columns are added to additional columns. Here is an augmented matrix, with three columns making up an identity matrix.

$$\begin{bmatrix} 3 & 6 & 5 & | & 1 & 0 & 0 \\ 2 & 8 & 9 & | & 0 & 1 & 0 \\ 4 & 7 & 10 & | & 0 & 0 & 1 \end{bmatrix}$$

vii. Using Matrices with Systems of Equations

Augmented matrices are frequently used to represent systems of equations. On one side of the augmented matrix, you can have the coefficients of the variables. On the other side, you have the constants.

For instance, the matrix $\begin{bmatrix} 5 & | & 20 \end{bmatrix}$ could represent the equation $5x = 20$. Likewise, the two–variable equation $3x - 4y = 16$ could be represented $\begin{bmatrix} 3 & -4 & | & 16 \end{bmatrix}$

To represent a system of equations, you would use multiple rows. You could use the following to represent the system consisting of $5x + 6y = 62$ and $7x + 9y = 91$:

$$\begin{array}{cc|c} 5 & 6 & 62 \\ 7 & 9 & 91 \end{array}$$

The first column includes the coefficients of the terms with the variable x, and the second column includes the coefficients of the terms with the variable y. The third column, which is on the other side of the augmented matrix, consists of the constants found on the right side of each equation.

To solve this system, you can perform row operations to get an identity matrix on the left side. The resulting numbers on the right side represent the values of the variables. If you did that here, you would get $\begin{array}{cc|c} 1 & 0 & 4 \\ 0 & 1 & 7 \end{array}$ which

indicates that $x = 4$ and $y = 7$, since the first row represents the variable x, and the second row represents the variable y.

Steps You Need to Remember

While some matrix operations, such as addition and subtraction, are rather straightforward, multiplication is more complicated. We'll explain all of those operations below. After that, we'll explain row operations, and show how you can use them to find inverses and solve systems of equations.

i. Adding and Subtracting

To add two matrices, add the corresponding elements. Remember that the matrices must be of the same size. So, the sum of the matrices

$$\begin{bmatrix} a & b \\ c & d \end{bmatrix} \text{ and } \begin{bmatrix} e & f \\ g & h \end{bmatrix} \text{ is } \begin{bmatrix} (a+e) & (b+f) \\ (c+g) & (d+h) \end{bmatrix}$$

Likewise, $\begin{bmatrix} a & b \\ c & d \end{bmatrix} - \begin{bmatrix} e & f \\ g & h \end{bmatrix} = \begin{bmatrix} (a-e) & (b-f) \\ (c-g) & (d-h) \end{bmatrix}$

ii. Multiplying

Take two matrices, *A* and *B*. You can multiply *A* by *B* if the number of columns in *A* is the same as the number of rows in *B*. So, *A* and *B* have a product if *A* is 2 x 2 and *B* is 2 x 3, but not if *A* is 2 x 2 and *B* is 3 x 2.

The product of *A* and *B*, which we'll call *AB*, has the same number of rows as *A* and the same number of columns as *B*. If *A* is 4 x 3 and *B* is 3 x 5, then *AB* is 4 x 5.

Each element in *AB* is the sum of the products of the elements of a row of *A* and a column of *B*. Take the element in the second row and third column of *AB*. That element is the sum of the product of the corresponding elements in the second row of *A* and the third column of *B*.

$\begin{bmatrix} \# & \# & \# \\ 2 & 3 & 4 \\ \# & \# & \# \end{bmatrix} \times \begin{bmatrix} \# & \# & 5 \\ \# & \# & 6 \\ \# & \# & 7 \end{bmatrix} = \begin{bmatrix} \# & \# & \# \\ \# & \# & 56 \\ \# & \# & \# \end{bmatrix}$, since

$2(5) + 3(6) + 4(7) = 10 + 18 + 28 = 56$.

We multiplied the first element in the row by the first element in the column, the second element in the row by the second element in the column, and the third element in the row by the third element in the column, and added the products to get 56.

In general, you get the element in the ith row and j^{th} column of *AB* by combining the ith row of *A* and the jth column of *B* in this way. You would find that you don't have a complete set of corresponding elements unless the number of columns in *A* and the number of rows in *B* are the same.

iii. Row Operations

The basic operations you can perform on rows are:

- Interchanging any two rows
- Multiplying or dividing a row by any number other than zero
- Adding a multiple of a row to another row

These operations can be combined. You can perform any or all of them on any row, as many times as you need.

When we perform row operations here, we'll use arrows to indicate where we're working, and we'll use the terms r_1, r_2, and so on, to indicate the rows involved. We'll also use algebraic expressions, with the row names as terms to describe the operations.

Here, we are tripling the second row of a matrix.

$$\begin{bmatrix} 4 & -2 & -1 \\ 8 & -5 & 4 \\ 7 & 3 & -3 \end{bmatrix} \xrightarrow{3r_2} \begin{bmatrix} 4 & -2 & -1 \\ 3(8) & 3(-5) & 3(4) \\ 7 & 3 & -3 \end{bmatrix} = \begin{bmatrix} 4 & -2 & -1 \\ 24 & -15 & 12 \\ 7 & 3 & -3 \end{bmatrix}$$

Here, we are adding twice the first row of a matrix to the second row:

$$\begin{bmatrix} 3 & 4 \\ 1 & 2 \end{bmatrix} \xrightarrow{r_2 + 2r_1} \begin{bmatrix} 3 & 4 \\ 1 + 2(3) & 2 + 2(4) \end{bmatrix} = \begin{bmatrix} 3 & 4 \\ 7 & 10 \end{bmatrix}$$

iv.   Finding the Inverse of a Matrix

In the **Concepts to Help You** section, we discussed augmented matrices. You can use them to find the inverses of matrices. Suppose you need to find the inverse of $\begin{bmatrix} 2 & -2 \\ 0 & 4 \end{bmatrix}$

Start by setting up an augmented matrix with the elements of that matrix on the left side, and the elements of an identity matrix on the right side:

$$\begin{bmatrix} 2 & -2 & | & 1 & 0 \\ 0 & 4 & | & 0 & 1 \end{bmatrix}$$

Next, you would perform certain row operations in order to get an augmented matrix with the elements of an identity matrix on the left side.

$$\begin{bmatrix} 2 & -2 & | & 1 & 0 \\ 0 & 4 & | & 0 & 1 \end{bmatrix} \rightarrow \begin{bmatrix} 1 & 0 & | & 0.5 & 0.25 \\ 0 & 1 & | & 0 & 0.25 \end{bmatrix}$$

Performing the right row operations on $\begin{bmatrix} 2 & -2 & | & 1 & 0 \\ 0 & 4 & | & 0 & 1 \end{bmatrix}$ would get you $\begin{bmatrix} 1 & 0 & | & 0.5 & 0.25 \\ 0 & 1 & | & 0 & 0.25 \end{bmatrix}$. The right side of this new augmented matrix is the inverse. So the inverse of $\begin{bmatrix} 2 & -2 \\ 0 & 4 \end{bmatrix}$ is $\begin{bmatrix} 0.5 & 0.25 \\ 0 & 0.25 \end{bmatrix}$. You could check this by finding that the product of these two is an identity matrix.

V. Solving a Linear System with Matrices

Solving a system of equations involves much of the same kind of work as finding the inverse of a matrix. When you are given a system of linear equations, you'll first need to set up an augmented matrix to represent it. To do this, you need to have all of the variable terms on the left side of each equation, and a constant on the right side. If the equations are not given in such a form, you need to work with them until you have them so.

Make sure that the terms in each equation are in the same order. If your system uses the variables x, y, and z, then you need to make sure that the x term, y term, and z term appear in the same order in each. Each column of the augmented matrix must include the coefficients for a single variable.

Once you've set up the augmented matrix properly, it's time to perform row operations. Your goal is to make the left side an identity matrix. What you get as a result on the right side of the augmented matrix are the values in the solution set of the linear system.

STEP-BY-STEP ILLUSTRATION OF THE FIVE MOST COMMON QUESTION TYPES

It's time once again to walk through several questions. The first two cover basic matrix operations. The third question involves row operations, a concept that is also essential to the last two questions.

Question 1: Adding Matrices

What is the sum of $\begin{bmatrix} 4 & -6 & 8 \\ -9 & 1 & -3 \end{bmatrix}$ and $\begin{bmatrix} -5 & -7 & 5 \\ 12 & -4 & -6 \end{bmatrix}$?

(A) $\begin{bmatrix} -9 & -13 & 13 \\ -3 & -5 & -9 \end{bmatrix}$ 　　(B) $\begin{bmatrix} -1 & -13 & 13 \\ -3 & -3 & 3 \end{bmatrix}$

(C) $\begin{bmatrix} -1 & -13 & 13 \\ 3 & -3 & -9 \end{bmatrix}$ 　　(D) $\begin{bmatrix} -1 & -13 & 13 \\ -14 & -5 & -9 \end{bmatrix}$

(E) $\begin{bmatrix} 1 & -1 & 13 \\ 3 & -5 & -9 \end{bmatrix}$

The sum of the two matrices is the matrix containing the sums of the corresponding elements:

$$\begin{bmatrix} 4 & -6 & 8 \\ -9 & 1 & -3 \end{bmatrix} + \begin{bmatrix} -5 & -7 & 5 \\ 12 & -4 & -6 \end{bmatrix} =$$

$$\begin{bmatrix} (4 + (-5)) & (-6 + (-7)) & (8 + 5) \\ (-9 + 12) & (1 + (-4)) & (-3 + (-6)) \end{bmatrix} =$$

$$\begin{bmatrix} -1 & -13 & 13 \\ 3 & -3 & -9 \end{bmatrix}$$

Choice (C) is the correct answer. Be very careful about matching corresponding elements. In choice (D), the value of -14 in the lower left element is the sum of the -9 in the first matrix and the -5 in the second.

Question 2: Multiplying Matrices

$$\begin{bmatrix} 4 & 1 & -2 \\ 3 & -1 & 2 \end{bmatrix} \times \begin{bmatrix} 5 & 7 \\ -4 & -6 \\ -3 & 4 \end{bmatrix} =$$

(A) $\begin{bmatrix} 10 & 14 \\ 13 & 35 \end{bmatrix}$ (B) $\begin{bmatrix} 10 & 14 \\ 25 & 35 \end{bmatrix}$

(C) $\begin{bmatrix} 22 & 13 \\ 14 & 35 \end{bmatrix}$ (D) $\begin{bmatrix} 22 & 14 \\ 13 & 35 \end{bmatrix}$

(E) $\begin{bmatrix} 22 & 26 \\ 13 & 23 \end{bmatrix}$

Since the first matrix has two rows and the second matrix has two columns, the product of the two is a 2 x 2 matrix. Let's set it up as follows:

$$\begin{bmatrix} P_{(1,1)} & P_{(1,2)} \\ P_{(2,1)} & P_{(2,2)} \end{bmatrix}$$

$P_{(i, j)}$ is the element in the ith row and j^{th} column of this matrix. We can get its value by combining the ith row of the first matrix with j^{th} column of the second matrix. We combine the row and the column by adding the products of the corresponding entries. So

$P_{(1, 1)}$ is the combination of the first row of $\begin{bmatrix} 4 & 1 & -2 \\ 3 & -1 & 2 \end{bmatrix}$ with the first

column $\begin{bmatrix} 5 & 7 \\ -4 & -6 \\ -3 & 4 \end{bmatrix}$

$$P_{(1, 1)} = 4(5) + 1(-4) + (-2)(-3) = 20 + (-4) + 6 = 22$$

So far, then, we know that $\begin{bmatrix} 4 & 1 & -2 \\ 3 & -1 & 2 \end{bmatrix} \times \begin{bmatrix} 5 & 7 \\ -4 & -6 \\ -3 & 4 \end{bmatrix} = \begin{bmatrix} 22 & P_{(1,2)} \\ P_{(2,1)} & P_{(2,2)} \end{bmatrix}$

Let's combine the rest of the pairs of rows and columns:

$$P_{(1, 2)} = 4(7) + 1(-6) + (-2)(4) = 28 + (-6) + (-8) = 14$$
$$P_{(2, 1)} = 3(5) + (-1)(-4) + 2(-3) = 15 + 4 + (-6) = 13$$

$$P_{(2,2)} = 3(7) + (-1)(-6) + 2(4) = 21 + 6 + 8 = 35$$

So, $\begin{bmatrix} P_{(1,1)} & P_{(1,2)} \\ P_{(2,1)} & P_{(2,2)} \end{bmatrix} = \begin{bmatrix} 22 & 14 \\ 13 & 35 \end{bmatrix}$ and **choice (D) is the correct answer.**

Choice (A) is the result of multiplying –4 by 4 instead of by 1 in getting the value of $P_{(1,\,1)}$.

Choice (C) is the product of $\begin{bmatrix} 5 & -4 & -3 \\ 7 & -6 & 4 \end{bmatrix}$ and $\begin{bmatrix} 4 & 3 \\ 1 & -1 \\ -2 & 2 \end{bmatrix}$

Question 3: Row Operations

Look at the matrix below.

$$\begin{bmatrix} 5 & -6 & 2 \\ -3 & 7 & 8 \\ 6 & 4 & -1 \end{bmatrix}$$

Which matrix do you get by doubling the first row, and adding three times the third row to the second row?

(A) $\begin{bmatrix} 10 & -6 & 2 \\ 15 & 19 & 5 \\ 6 & 4 & -1 \end{bmatrix}$ (B) $\begin{bmatrix} 10 & -6 & 2 \\ 15 & 19 & 5 \\ -9 & 21 & 24 \end{bmatrix}$

(C) $\begin{bmatrix} 10 & -12 & 4 \\ 18 & 12 & -3 \\ 6 & 4 & -1 \end{bmatrix}$ (D) $\begin{bmatrix} 10 & -12 & 4 \\ -3 & 7 & 8 \\ -3 & 25 & 23 \end{bmatrix}$

(E) $\begin{bmatrix} 10 & -12 & 4 \\ 15 & 19 & 5 \\ 6 & 4 & -1 \end{bmatrix}$

To find out which matrix is the result of performing the given operations, carry them out one at a time.

First, doubling the first row gives you the following:

$$\begin{bmatrix} 5 & -6 & 2 \\ -3 & 7 & 8 \\ 6 & 4 & -1 \end{bmatrix} \xrightarrow{2r_1} \begin{bmatrix} 5(2) & -6(2) & 2(2) \\ -3 & 7 & 8 \\ 6 & 4 & -1 \end{bmatrix} = \begin{bmatrix} 10 & -12 & 4 \\ -3 & 7 & 8 \\ 6 & 4 & -1 \end{bmatrix}$$

Note that in choices (A) and (B), only the first element in the row are doubled. The other two stay the same.

Second, add three times the third row to the second row. The elements in the third row are 6, 4, and –1, so you need to add 18, 12, and –3 to the corresponding elements in row 2:

$$\begin{bmatrix} 10 & -12 & 4 \\ -3 & 7 & 8 \\ 6 & 4 & -1 \end{bmatrix} \xrightarrow{3r_3 + r_2} \begin{bmatrix} 10 & -12 & 4 \\ -3+18 & 7+12 & 8+(-3) \\ 6 & 4 & -1 \end{bmatrix} = \begin{bmatrix} 10 & -12 & 4 \\ 15 & 19 & 5 \\ 6 & 4 & -1 \end{bmatrix}$$

Both operations have been performed. **Choice (E) is the correct answer.**

Even though the third row is used in the second operation (performed on the second row), no operation is actually performed on the third row. That row remains unchanged. Choice (D) gets that row operation backward, however, adding triple the second row to the third. Choice (C) also results from performing the second operation incorrectly. There, it replaces the second row with triple the third row, instead of adding triple the third row to the values already in the second row.

Question 4: Matrix Inversion

What is the inverse of $\begin{bmatrix} 4 & 6 \\ 1 & 2 \end{bmatrix}$?

(A) $\begin{bmatrix} -4 & -6 \\ -1 & -2 \end{bmatrix}$

(B) $\begin{bmatrix} 0.125 & 0.75 \\ 0.25 & -0.5 \end{bmatrix}$

(C) $\begin{bmatrix} 0.25 & -0.75 \\ 0 & 0.5 \end{bmatrix}$

(D) $\begin{bmatrix} 0.25 & 0.75 \\ 1 & 0.5 \end{bmatrix}$

(E) $\begin{bmatrix} 1 & -3 \\ -0.5 & 2 \end{bmatrix}$

To find the inverse of the matrix, create an augmented matrix with the identity matrix:

$$\begin{bmatrix} 4 & 6 & | & 1 & 0 \\ 1 & 2 & | & 0 & 1 \end{bmatrix}$$

Now, perform row operations on the augmented matrix, so as to change the first two columns into an identity matrix. This may require some trial and error; it can be difficult to get the right result. Just remember that you have to carry out operations on the entire row of the augmented matrix.

To change the first row, with the elements 4 and 6, to one with the elements 1 and 0, you can multiply the second row by three, and subtract the results from the first row:

$$\begin{bmatrix} 4 & 6 & | & 1 & 0 \\ 1 & 2 & | & 0 & 1 \end{bmatrix} \xrightarrow{r_1-3r_2} \begin{bmatrix} 4-3 & 6-6 & | & 1-0 & 0-3 \\ 1 & 2 & | & 0 & 1 \end{bmatrix} = \begin{bmatrix} 1 & 0 & | & 1 & -3 \\ 1 & 2 & | & 0 & 1 \end{bmatrix}$$

We subtracted –3(1) from 4, –3(2) from 6, –3(0) from 1, and –3(1) from 0. That gives us the elements 1 and 0 on the left side of the augmented matrix, and 1 and –3 on the right side.

Now we need to get the bottom row into shape. To get the elements 0 and 1 from 1 and 2, you can subtract the first row from the second row, and multiply the difference by 0.5 (i.e., divide the difference by 2).

$$\begin{bmatrix} 1 & 0 & | & 1 & -3 \\ 1 & 2 & | & 0 & 1 \end{bmatrix} \xrightarrow{(r_2-r_1)\times0.5} \begin{bmatrix} 1 & 0 & | & 1 & -3 \\ (1-1)\times0.5 & (2-0)\times0.5 & | & (0-1)\times0.5 & (1-(-3))\times0.5 \end{bmatrix}$$

$$= \begin{bmatrix} 1 & 0 & | & 1 & -3 \\ 0 & 1 & | & -0.5 & 2 \end{bmatrix}$$

Choice (E) is the correct answer. If you worked on the solution on your own, you might have gotten the correct answer by a different route. The row operations we performed represent just one way of changing the matrix into an identity matrix.

Choice (A) includes the additive inverses of the elements of the original matrix. The product of the two would not be the identity matrix, however.

Choice (B) is the inverse $\begin{bmatrix} 4 & 6 \\ 2 & 1 \end{bmatrix}$ rather than $\begin{bmatrix} 4 & 6 \\ 1 & 2 \end{bmatrix}$. Choice (C) is the inverse of $\begin{bmatrix} 4 & 6 \\ 0 & 2 \end{bmatrix}$.

Question 5: Using Matrices to Solve Systems of Equations

If $2x - 3y - 3z = 0$, $3x + 2y + 2z = 13$, and $x - 2y + z = 20$, then

(A) $x = -2$, $y = 5$, $z = 7$

(B) $x = 2$, $y = 3$, $z = 7$

(C) $x = 3$, $y = -5$, $z = 7$

(D) $x = 3$, $y = 5$, $z = -9$

(E) $x = 3$, $y = -7$, $z = 9$

You can solve this system of equations by performing row operations on an augmented matrix. Each row represents one of the equations. Each of the three columns on one side of the augmented matrix will represent one variable in the system. The first column will have the coefficients of the x terms, the second will have the coefficients of the y terms, and so on. The fourth column, the one on the other side of the augmented matrix, gives the numbers on the right side of each equation. So, the augmented matrix representing the system

$$2x - 3y - 3z = 0$$
$$3x + 2y + 2z = 13$$
$$x - 2y + z = 20$$

is $\left[\begin{array}{ccc|c} 2 & -3 & -3 & 0 \\ 3 & 2 & 2 & 13 \\ 1 & -2 & 1 & 20 \end{array}\right]$

The goal here is to perform row operations to make the left side of the augmented matrix an identity matrix. The resulting numbers on the right side will represent the values of each variable.

We can begin by multiplying the first row by 2 and adding three times the second row to it:

$$\left[\begin{array}{ccc|c} 2 & -3 & -3 & 0 \\ 3 & 2 & 2 & 13 \\ 1 & -2 & 1 & 20 \end{array}\right] \xrightarrow{2r_1 + 3r_2} \left[\begin{array}{ccc|c} 2(2)+3(3) & 2(-3)+3(2) & 2(-3)+3(2) & 2(0)+3(13) \\ 3 & 2 & 2 & 13 \\ 1 & -2 & 1 & 20 \end{array}\right]$$

$$= \left[\begin{array}{ccc|c} 13 & 0 & 0 & 39 \\ 3 & 2 & 2 & 13 \\ 1 & -2 & 1 & 20 \end{array}\right]$$

To finish up operations on this row, divide it by 13:

$$\begin{bmatrix} 13 & 0 & 0 & | & 39 \\ 3 & 2 & 2 & | & 13 \\ 1 & -2 & 1 & | & 20 \end{bmatrix} \xrightarrow{\;r_1 \div 13\;} \begin{bmatrix} 1 & 0 & 0 & | & 3 \\ 3 & 2 & 2 & | & 13 \\ 1 & -2 & 1 & | & 20 \end{bmatrix}$$

Next, we can take care of the third row by adding the second row, subtracting four times the first row, and then dividing the row by 3:

$$\begin{bmatrix} 1 & 0 & 0 & | & 3 \\ 3 & 2 & 2 & | & 13 \\ 1 & -2 & 1 & | & 20 \end{bmatrix} \xrightarrow{\;r_3 + r_2 - 4r_3\;} \begin{bmatrix} 1 & 0 & 0 & | & 3 \\ 3 & 2 & 2 & | & 13 \\ 1+3-4(1) & -2+2-4(0) & 1+2-4(0) & | & 20+13-4(3) \end{bmatrix}$$

$$= \begin{bmatrix} 1 & 0 & 0 & | & 3 \\ 3 & 2 & 2 & | & 13 \\ 0 & 0 & 3 & | & 21 \end{bmatrix} \xrightarrow{\;r_3 \div 3\;} \begin{bmatrix} 1 & 0 & 0 & | & 3 \\ 3 & 2 & 2 & | & 13 \\ 0 & 0 & 1 & | & 7 \end{bmatrix}$$

Finally, we can perform several operations on the second row:

$$\begin{bmatrix} 1 & 0 & 0 & | & 3 \\ 3 & 2 & 2 & | & 13 \\ 0 & 0 & 1 & | & 7 \end{bmatrix} \xrightarrow{\;r_2 - 3r_1 - 2r_3\;} \begin{bmatrix} 1 & 0 & 0 & | & 3 \\ 3-3(1)-2(0) & 2-3(0)-2(0) & 2-3(0)-2(1) & | & 13-3(3)-2(7) \\ 0 & 0 & 1 & | & 7 \end{bmatrix}$$

$$= \begin{bmatrix} 1 & 0 & 0 & | & 3 \\ 0 & 2 & 0 & | & -10 \\ 0 & 0 & 1 & | & 7 \end{bmatrix} \xrightarrow{\;r_2 \div 2\;} \begin{bmatrix} 1 & 0 & 0 & | & 3 \\ 0 & 1 & 0 & | & -5 \\ 0 & 0 & 1 & | & 7 \end{bmatrix}$$

We now have the augmented matrix $\begin{bmatrix} 1 & 0 & 0 & | & 3 \\ 0 & 1 & 0 & | & -5 \\ 0 & 0 & 1 & | & 7 \end{bmatrix}$

Since the first column represents x, the second column y, and the third one z, the solution is $x = 3$, $y = -5$, and $z = 7$. By performing row operations, we derived those equations from the three given by the question. **(C) is the correct answer choice.**

CHAPTER QUIZ

1. $\begin{bmatrix} 4 & 20 \\ -4 & -7 \\ -6 & 8 \end{bmatrix} - \begin{bmatrix} 7 & 11 \\ -8 & 2 \\ 9 & 12 \end{bmatrix} =$

    (A) $\begin{bmatrix} -3 & 9 \\ -12 & -5 \\ -3 & -4 \end{bmatrix}$
    (B) $\begin{bmatrix} -3 & 9 \\ 4 & -9 \\ -15 & -4 \end{bmatrix}$
    (C) $\begin{bmatrix} -3 & 9 \\ 4 & -9 \\ -3 & -4 \end{bmatrix}$

    (D) $\begin{bmatrix} 3 & 9 \\ 4 & -5 \\ -3 & -4 \end{bmatrix}$
    (E) $\begin{bmatrix} 3 & 9 \\ 4 & -5 \\ -15 & -4 \end{bmatrix}$

2. If $\begin{bmatrix} 8 & -4 & 7 \\ 9 & 6 & -1 \end{bmatrix} + A = \begin{bmatrix} 12 & -6 & 4 \\ 8 & 7 & 0 \end{bmatrix}$ then $A =$

    (A) $\begin{bmatrix} -4 & -3 & -2 \\ -1 & 1 & 1 \end{bmatrix}$
    (B) $\begin{bmatrix} -4 & 2 & -3 \\ 1 & -1 & 1 \end{bmatrix}$
    (C) $\begin{bmatrix} 4 & -2 & -3 \\ -1 & -1 & -1 \end{bmatrix}$

    (D) $\begin{bmatrix} 4 & -3 & -2 \\ -1 & -1 & -1 \end{bmatrix}$
    (E) $\begin{bmatrix} 4 & -2 & -3 \\ -1 & 1 & 1 \end{bmatrix}$

3. If $A = \begin{bmatrix} 7 & 8 \\ 5 & 2 \end{bmatrix}$ and $B = \begin{bmatrix} -1 & 4 & -3 & 5 \\ 3 & -6 & -2 & -4 \end{bmatrix}$ then $A \times B =$

    (A) $\begin{bmatrix} 1 & -20 & -37 & 3 \\ -3 & 8 & -19 & 17 \end{bmatrix}$
    (B) $\begin{bmatrix} 17 & -20 & -37 & 3 \\ 1 & 8 & -19 & 17 \end{bmatrix}$

    (C) $\begin{bmatrix} 17 & -76 & -37 & 3 \\ 1 & 32 & -19 & 17 \end{bmatrix}$
    (D) $\begin{bmatrix} 31 & -20 & -37 & 3 \\ 11 & 8 & -19 & 17 \end{bmatrix}$

    (E) $\begin{bmatrix} 31 & -20 & 37 & 3 \\ 11 & 8 & 19 & 17 \end{bmatrix}$

4. $\begin{bmatrix} 5 & 3 & 7 \\ 6 & 2 & -8 \\ 1 & 4 & -2 \end{bmatrix} \times \begin{bmatrix} -4 & 10 \\ 9 & -5 \\ -1 & -2 \end{bmatrix} =$

(A) $\begin{bmatrix} 0 & 21 \\ 2 & 66 \\ 34 & -6 \end{bmatrix}$ (B) $\begin{bmatrix} 0 & 51 \\ 2 & 86 \\ 34 & 34 \end{bmatrix}$ (C) $\begin{bmatrix} 33 & 2 & -98 \\ 18 & 12 & 114 \end{bmatrix}$

(D) $\begin{bmatrix} 40 & 21 \\ 50 & 66 \\ 42 & -6 \end{bmatrix}$ (E) $\begin{bmatrix} 73 & 26 & -42 \\ 18 & 12 & 114 \end{bmatrix}$

5. Which matrix results from doubling the first row of $\begin{bmatrix} 9 & -3 & 5 \\ -6 & 8 & -2 \end{bmatrix}$ and adding the second row to it, and then dividing the second row by 2?

(A) $\begin{bmatrix} 9 & -3 & 5 \\ 6 & 1 & 4 \end{bmatrix}$ (B) $\begin{bmatrix} 9 & -3 & 5 \\ 12 & 2 & 8 \end{bmatrix}$ (C) $\begin{bmatrix} 12 & 2 & 8 \\ -3 & 4 & -1 \end{bmatrix}$

(D) $\begin{bmatrix} 15 & -2 & 9 \\ -3 & 4 & -1 \end{bmatrix}$ (E) $\begin{bmatrix} 18 & -6 & 10 \\ -3 & 4 & -1 \end{bmatrix}$

6. Which of these sets of row operations could be performed on $\begin{bmatrix} 4 & 6 & -3 \\ 5 & -2 & 8 \end{bmatrix}$ to get $\begin{bmatrix} 13 & 10 & 2 \\ 8 & 12 & -6 \end{bmatrix}$?

(A) Adding the first row to the second and interchanging the rows

(B) Adding twice the first row to the second and adding the second row to the first

(C) Adding twice the first row to the second, adding the second row to the first, and interchanging the rows

(D) Adding twice the second row to the first, doubling the second row, and interchanging the rows

(E) Adding twice the first row to the second, doubling the first row, and interchanging the rows

7.   What is the inverse of $\begin{bmatrix} 4 & -2 \\ -1 & 3 \end{bmatrix}$?

(A) $\begin{bmatrix} 0.1 & 0.2 \\ 0.4 & 0.3 \end{bmatrix}$ (B) $\begin{bmatrix} 0.2 & 0.1 \\ 0.3 & 0.4 \end{bmatrix}$ (C) $\begin{bmatrix} 0.3 & 0.2 \\ 0.1 & 0.4 \end{bmatrix}$

(D) $\begin{bmatrix} 0.4 & 0.2 \\ 0.1 & 0.3 \end{bmatrix}$ (E) $\begin{bmatrix} 0.4 & 0.1 \\ 0.2 & 0.3 \end{bmatrix}$

8. If $A = \begin{bmatrix} -1 & 0 & 1 \\ 2 & -1 & -2 \\ -2 & 1 & 3 \end{bmatrix}$ then $A^{-1} =$

(A) $\begin{bmatrix} -3 & 1 & 1 \\ -2 & 1 & 0 \\ 0 & 1 & 1 \end{bmatrix}$ (B) $\begin{bmatrix} -2 & 1 & 3 \\ -1 & -1 & 0 \\ 0 & 1 & 1 \end{bmatrix}$ (C) $\begin{bmatrix} -2 & 1 & 1 \\ 2 & -1 & 0 \\ 0 & 1 & -3 \end{bmatrix}$

(D) $\begin{bmatrix} -1 & 1 & 1 \\ -2 & -1 & 0 \\ 0 & 1 & 1 \end{bmatrix}$ (E) $\begin{bmatrix} -1 & 1 & 1 \\ -1 & -2 & 0 \\ 0 & 2 & 1 \end{bmatrix}$

9. Which matrix represents the solution set of $3x + 2y = 22$ and $-5x + 7y = 46$?

(A) $\begin{bmatrix} 1 & 0 & | & -8 \\ 0 & 1 & | & 2 \end{bmatrix}$ (B) $\begin{bmatrix} 1 & 0 & | & -8 \\ 0 & 1 & | & 4 \end{bmatrix}$ (C) $\begin{bmatrix} 1 & 0 & | & 2 \\ 0 & 1 & | & 8 \end{bmatrix}$

(D) $\begin{bmatrix} 1 & 0 & | & 4 \\ 0 & 1 & | & 8 \end{bmatrix}$ (E) $\begin{bmatrix} 1 & 0 & | & 8 \\ 0 & 1 & | & -4 \end{bmatrix}$

10. Look at the system of equations below.

$$2a + 2b - c = -5$$
$$3a - b + 3c = 9$$
$$-a + 3b - 2c = 4$$

What are the values of the variables?

(A) $a = -6$, $b = -4$, $c = 9$

(B) $a = -4$, $b = -6$, $c = 9$

(C) $a = -4$, $b = 6$, $c = 9$

(D) $a = 4$, $b = -6$, $c = -9$

(E) $a = 6$, $b = 4$, $c = -9$

ANSWER EXPLANATIONS

1. B

$$\begin{bmatrix} 4 & 20 \\ -4 & -7 \\ -6 & 8 \end{bmatrix} - \begin{bmatrix} 7 & 11 \\ -8 & 2 \\ 9 & 12 \end{bmatrix} = \begin{bmatrix} (4-7) & (20-11) \\ (-4-(-8)) & (-7-2) \\ (-6-9) & (8-12) \end{bmatrix} = \begin{bmatrix} -3 & 9 \\ 4 & -9 \\ -15 & -4 \end{bmatrix}$$

2. E

You can use the rules of matrix addition to recognize that this question requires you to solve several basic equations. Since A is added to

$\begin{bmatrix} 8 & -4 & 7 \\ 9 & 6 & -1 \end{bmatrix}$, it must also be a 2 x 3 matrix. Each of the six elements in A is

unknown, so you can treat them as variables. Using the letters u, v, w, x, y, z,

$$A = \begin{bmatrix} u & v & w \\ x & y & z \end{bmatrix}$$

$$\begin{bmatrix} 8 & -4 & 7 \\ 9 & 6 & -1 \end{bmatrix} + A = \begin{bmatrix} 8 & -4 & 7 \\ 9 & 6 & -1 \end{bmatrix} + \begin{bmatrix} u & v & w \\ x & y & z \end{bmatrix} = \begin{bmatrix} 8+u & -4+v & 7+w \\ 9+x & 6+y & -1+z \end{bmatrix} = \begin{bmatrix} 12 & -6 & 4 \\ 8 & 7 & 0 \end{bmatrix}$$

So, there are six equations to solve:

$8 + u = 12 \rightarrow u = 4$

$-4 + v = -6 \rightarrow v = -2$

$7 + w = 4 \rightarrow w = -3$

$9 + x = 8 \rightarrow x = -1$

$6 + y = 7 \rightarrow y = 1$

$-1 + z = 0 \rightarrow z = 1$

Plug these solutions into A:

$$A = \begin{bmatrix} u & v & w \\ x & y & z \end{bmatrix} = \begin{bmatrix} 4 & -2 & -3 \\ -1 & 1 & 1 \end{bmatrix}$$

3. B

$$\begin{bmatrix} 7 & 8 \\ 5 & 2 \end{bmatrix} \times \begin{bmatrix} -1 & 4 & -3 & 5 \\ 3 & -6 & -2 & -4 \end{bmatrix} = \begin{bmatrix} P_{(1,1)} & P_{(1,2)} & P_{(1,3)} & P_{(1,4)} \\ P_{(2,1)} & P_{(2,2)} & P_{(2,3)} & P_{(2,4)} \end{bmatrix}$$

$P_{(1, 1)} = 7(-1) + 8(3) = -7 + 24 = 17$

$P_{(1, 2)} = 7(4) + 8(-6) = 28 + (-48) = -20$

$P_{(1, 3)} = 7(-3) + 8(-2) = -21 + (-16) = -37$

$P_{(1, 4)} = 7(5) + 8(-4) = 35 + (-32) = 3$

$P_{(2, 1)} = 5(-1) + 2(3) = -5 + 6 = 1$

$P_{(2, 2)} = 5(4) + 2(-6) = 20 + (-12) = 8$

$P_{(2, 3)} = 5(-3) + 2(-2) = -15 + (-4) = -19$

$P_{(2, 4)} = 5(5) + 2(-4) = 25 + (-8) = 17$

So $\begin{bmatrix} P_{(1,1)} & P_{(1,2)} & P_{(1,3)} & P_{(1,4)} \\ P_{(2,1)} & P_{(2,2)} & P_{(2,3)} & P_{(2,4)} \end{bmatrix} = \begin{bmatrix} 17 & -20 & -37 & 3 \\ 1 & 8 & -19 & 17 \end{bmatrix}$

4. A

$$\begin{bmatrix} 5 & 3 & 7 \\ 6 & 2 & -8 \\ 1 & 4 & -2 \end{bmatrix} \times \begin{bmatrix} -4 & 10 \\ 9 & -5 \\ -1 & -2 \end{bmatrix} = \begin{bmatrix} P_{(1,1)} & P_{(1,2)} \\ P_{(2,1)} & P_{(2,2)} \\ P_{(3,1)} & P_{(3,2)} \end{bmatrix}$$

$P_{(1, 1)} = 5(-4) + 3(9) + 7(-1) = -20 + 27 + (-7) = 0$

$P_{(1, 2)} = 5(10) + 3(-5) + 7(-2) = 50 + (-15) + (-14) = 21$

$P_{(2, 1)} = 6(-4) + 2(9) + (-8)(-1) = -24 + 18 + 8 = 2$

$P_{(2, 2)} = 6(10) + 2(-5) + (-8)(-2) = 60 + (-10) + 16 = 66$

$P_{(3, 1)} = 1(-4) + 4(9) + (-2)(-1) = -4 + 36 + 2 = 34$

$P_{(3, 2)} = 1(10) + 4(-5) + (-2)(-2) = 10 + (-20) + 4 = -6$

So $\begin{bmatrix} P_{(1,1)} & P_{(1,2)} \\ P_{(2,1)} & P_{(2,2)} \\ P_{(3,1)} & P_{(3,2)} \end{bmatrix} = \begin{bmatrix} 0 & 21 \\ 2 & 66 \\ 34 & -6 \end{bmatrix}$

Choice (C) is the product of $\begin{bmatrix} -4 & 9 & -1 \\ 10 & -5 & -2 \end{bmatrix}$ and $\begin{bmatrix} 5 & 3 & 7 \\ 6 & 2 & -8 \\ 1 & 4 & -2 \end{bmatrix}$

5. C

$$\begin{bmatrix} 9 & -3 & 5 \\ -6 & 8 & -2 \end{bmatrix} \xrightarrow[\;r_2 + 2\;]{2r_1 + r_2} \begin{bmatrix} 2(9) + (-6) & 2(-3) + 8 & 2(5) + (-2) \\ -6 + 2 & 8 + 2 & -2 + 2 \end{bmatrix}$$

$$= \begin{bmatrix} 12 & 2 & 8 \\ -3 & 4 & -1 \end{bmatrix}$$

Choice (A) is the result of adding twice the first row to the second row, instead of adding the second row to twice the first. Choice (D) is the result of performing the row operations in the reverse order.

6. E

To identify the correct set of operations, you can perform each set on

$\begin{bmatrix} 4 & 6 & -3 \\ 5 & -2 & 8 \end{bmatrix}$ and see which one results in $\begin{bmatrix} 13 & 10 & 2 \\ 8 & 12 & -6 \end{bmatrix}$

Here is the result of adding twice the first row to the second and doubling the first row:

$$\begin{bmatrix} 4 & 6 & -3 \\ 5 & -2 & 8 \end{bmatrix} \xrightarrow[\;2r_1 + r_2\;]{2r_1} \begin{bmatrix} 2(4) & 2(6) & 2(-3) \\ 2(4) + 5 & 2(6) + (-2) & 2(-3) + 8 \end{bmatrix} = \begin{bmatrix} 8 & 12 & -6 \\ 13 & 10 & 2 \end{bmatrix}$$

Interchanging the new rows results in $\begin{bmatrix} 13 & 10 & 2 \\ 8 & 12 & -6 \end{bmatrix}$

Performing the operations in choice (A) results in $\begin{bmatrix} 9 & 4 & 5 \\ 4 & 6 & -3 \end{bmatrix}$

The operations in choice (B) result in $\begin{bmatrix} 9 & 4 & 5 \\ 13 & 10 & 2 \end{bmatrix}$ Choice (C) has these new rows interchanged.

The operations in choice (D) result in $\begin{bmatrix} 10 & -4 & 16 \\ 14 & 2 & 13 \end{bmatrix}$

7. C

Carry out operations on each row of the matrix to convert it to an identity matrix. By performing the same operations on an identity matrix, you can obtain the inverse.

$$\begin{bmatrix} 4 & -2 & | & 1 & 0 \\ -1 & 3 & | & 0 & 1 \end{bmatrix} \xrightarrow{(3r_1 + 2r_2) \div 10} \begin{bmatrix} 1 & 0 & | & 0.3 & 0.2 \\ -1 & 3 & | & 0 & 1 \end{bmatrix}$$

$$\begin{bmatrix} 1 & 0 & | & 0.3 & 0.2 \\ -1 & 3 & | & 0 & 1 \end{bmatrix} \xrightarrow{(r_1 + r_2) \div 3} \begin{bmatrix} 1 & 0 & | & 0.3 & 0.2 \\ 0 & 1 & | & 0.1 & 0.4 \end{bmatrix}$$

8. D

Remember that matrix A^{-1} is just the inverse of A. Perform row operations to change A into an identity matrix. The very same row operations, performed on the identity matrix, will give you A^{-1}.

$$\begin{bmatrix} -1 & 0 & 1 & | & 1 & 0 & 0 \\ 2 & -1 & -2 & | & 0 & 1 & 0 \\ -2 & 1 & 3 & | & 0 & 0 & 1 \end{bmatrix} \xrightarrow{-r_1 + r_2 + r_3} \begin{bmatrix} 1 & 0 & 0 & | & -1 & 1 & 1 \\ 2 & -1 & -2 & | & 0 & 1 & 0 \\ -2 & 1 & 3 & | & 0 & 0 & 1 \end{bmatrix}$$

$$\begin{bmatrix} 1 & 0 & 0 & | & -1 & 1 & 1 \\ 2 & -1 & -2 & | & 0 & 1 & 0 \\ -2 & 1 & 3 & | & 0 & 0 & 1 \end{bmatrix} \xrightarrow{r_2 + r_3} \begin{bmatrix} 1 & 0 & 0 & | & -1 & 1 & 1 \\ 2 & -1 & -2 & | & 0 & 1 & 0 \\ 0 & 0 & 1 & | & 0 & 1 & 1 \end{bmatrix}$$

$$\begin{bmatrix} 1 & 0 & 0 & | & -1 & 1 & 1 \\ 2 & -1 & -2 & | & 0 & 1 & 0 \\ 0 & 0 & 1 & | & 0 & 1 & 1 \end{bmatrix} \xrightarrow{-r_2 + 2r_1 - 2r_3} \begin{bmatrix} 1 & 0 & 0 & | & -1 & 1 & 1 \\ 0 & 1 & 0 & | & -2 & -1 & 0 \\ 0 & 0 & 1 & | & 0 & 1 & 1 \end{bmatrix}$$

9. C

This matrix represents the given system of equations:

$$\begin{bmatrix} 3 & 2 & | & 22 \\ -5 & 7 & | & 46 \end{bmatrix}$$

You can start by performing operations on the second row:

$$\begin{bmatrix} 3 & 2 & | & 22 \\ -5 & 7 & | & 46 \end{bmatrix} \xrightarrow{3r_2+5r_1} \begin{bmatrix} 3 & 2 & | & 22 \\ 3(-5)+5(3) & 3(7)+5(2) & | & 3(46)+5(22) \end{bmatrix} = \begin{bmatrix} 3 & 2 & | & 22 \\ 0 & 31 & | & 248 \end{bmatrix}$$

$$\begin{bmatrix} 3 & 2 & | & 22 \\ 0 & 31 & | & 248 \end{bmatrix} \xrightarrow{r_2 \div 31} \begin{bmatrix} 3 & 2 & | & 22 \\ 0 & 1 & | & 8 \end{bmatrix}$$

Now carry out operations on the first row:

$$\begin{bmatrix} 3 & 2 & | & 22 \\ 0 & 1 & | & 8 \end{bmatrix} \xrightarrow{r_1-2r_2} \begin{bmatrix} 3-2(0) & 2-2(1) & | & 22-2(8) \\ 0 & 1 & | & 8 \end{bmatrix} = \begin{bmatrix} 3 & 0 & | & 6 \\ 0 & 1 & | & 8 \end{bmatrix} \xrightarrow{r_1 \div 3} \begin{bmatrix} 1 & 0 & | & 2 \\ 0 & 1 & | & 8 \end{bmatrix}$$

Since the result is an augmented matrix with an identity matrix on the left side, it represents the solution set of the given system of equations.

10. C

Use this augmented matrix to represent the system of equations:

$$\begin{bmatrix} 2 & 2 & -1 & | & -5 \\ 3 & -1 & 3 & | & 9 \\ -1 & 3 & -2 & | & 4 \end{bmatrix}$$

Now perform row operations to get an identity matrix:

$$\begin{bmatrix} 2 & 2 & -1 & | & -5 \\ 3 & -1 & 3 & | & 9 \\ -1 & 3 & -2 & | & 4 \end{bmatrix} \xrightarrow{r_3+r_2-r_1} \begin{bmatrix} 2 & 2 & -1 & | & -5 \\ 3 & -1 & 3 & | & 9 \\ -1+3-2 & 3+(-1)-2 & -2+3-(-1) & | & 4+9-(-5) \end{bmatrix}$$

$$= \begin{bmatrix} 2 & 2 & -1 & | & -5 \\ 3 & -1 & 3 & | & 9 \\ 0 & 0 & 2 & | & 18 \end{bmatrix} \xrightarrow{r_3 \div 2} \begin{bmatrix} 2 & 2 & -1 & | & -5 \\ 3 & -1 & 3 & | & 9 \\ 0 & 0 & 1 & | & 9 \end{bmatrix}$$

$$\begin{bmatrix} 2 & 2 & -1 & | & -5 \\ 3 & -1 & 3 & | & 9 \\ 0 & 0 & 1 & | & 9 \end{bmatrix} \xrightarrow{r_1 + 2r_2 - 5r_3} \begin{bmatrix} 2 + 2(3) - 5(0) & 2 + 2(-1) - 5(0) & -1 + 2(3) - 5(1) & | & -5 + 2(9) - 5(9) \\ 3 & -1 & 3 & | & 9 \\ 0 & 0 & 1 & | & 9 \end{bmatrix}$$

$$= \begin{bmatrix} 8 & 0 & 0 & | & -32 \\ 3 & -1 & 3 & | & 9 \\ 0 & 0 & 1 & | & 9 \end{bmatrix} \xrightarrow{r_1 \div 8} \begin{bmatrix} 1 & 0 & 0 & | & -4 \\ 3 & -1 & 3 & | & 9 \\ 0 & 0 & 1 & | & 9 \end{bmatrix}$$

$$\begin{bmatrix} 1 & 0 & 0 & | & -4 \\ 3 & -1 & 3 & | & 9 \\ 0 & 0 & 1 & | & 9 \end{bmatrix} \xrightarrow{r_2 - 3r_1 - 3r_3} \begin{bmatrix} 1 & 0 & 0 & | & -4 \\ 3 - 3(1) - 3(0) & -1 - 3(0) - 3(0) & 3 - 3(0) - 3(1) & | & 9 - 3(-4) - 3(9) \\ 0 & 0 & 1 & | & 9 \end{bmatrix}$$

$$= \begin{bmatrix} 1 & 0 & 0 & | & -4 \\ 0 & -1 & 0 & | & -6 \\ 0 & 0 & 1 & | & 9 \end{bmatrix} \xrightarrow{-1 \times r} \begin{bmatrix} 1 & 0 & 0 & | & -4 \\ 0 & 1 & 0 & | & 6 \\ 0 & 0 & 1 & | & 9 \end{bmatrix}$$

Now we have an identity matrix on the left side of the augmented matrix. It indicates that $a = -4$, $b = 6$, and $c = 9$.

Conic Sections

WHAT ARE CONIC SECTIONS?

A *conic section* is a two-dimensional figure created by the intersection of a plane and a double cone. A double cone like the one shown below is literally two cones, making contact at each apex.

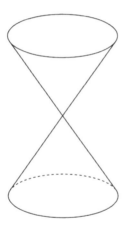

CONCEPTS TO HELP YOU

Depending on how a plane intersects a double cone, different kinds of conic sections can be created. The four we'll focus on here are the *circle*, the *ellipse*, the *parabola*, and the *hyperbola*. Two other conic sections, which don't usually get much attention, are the point and the pair of intersecting lines.

Each of the four conic sections we'll study has important properties. We'll see how they are represented algebraically and graphically.

i. The Double Cone

The individual cones are called *nappes*. Each nappe opens at the same angle, and each one actually extends infinitely. The double cone has an *axis*, a straight line that bisects the angle at which each nappe opens.

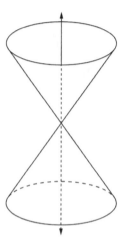

The circular areas at the top and bottom of the figure are really just cross–sections of the double cone. The double cone has no bases, unlike a three–dimensional solid.

ii. The Circle

The figure created by the intersection of the double cone and a plane perpendicular to the axis is a circle (unless the plane intersects the apexes).

A circle is defined as the set of points on a plane that are equally distant from a single point. That point is the center of the circle. The distance from the center to any point on the circle is the radius.

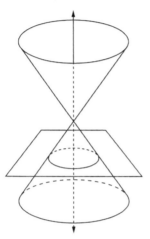

A circle on the coordinate plane with a center at the origin can be represented with the equation

$x^2 + y^2 = r^2$, where r is the radius of the circle.

So, a circle centered at the origin includes point P if the sum of the squares of its coordinates equals the square of the radius.

A circle with a center located at (h, k) and a radius r has the equation

$(x - h)^2 + (y - k)^2 = r^2$.

iii. The Ellipse

An ellipse is the figure created by a plane intersecting the nappe of a double cone at an angle such that it crosses the axis, without being perpendicular to it.

You can think of an ellipse as an oval, a circle that has been stretched horizontally or vertically. More technically, an ellipse can be defined in terms of focus points, or *foci*. An ellipse has two foci, and the sums of the distances of each point on the ellipse from each focus are the same.

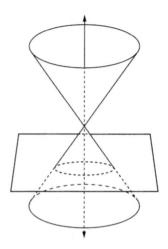

The curved figure below is an ellipse. The points F_1 and F_2 are the foci. Points D and E are on the ellipse.

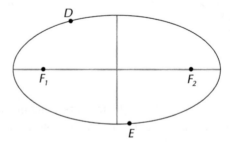

The sum of the distance between D and F_1 and the distance between D and F_2 equals the sum of the distance between E and F_1 and the distance between E and F_2. That relationship holds for any two points on the ellipse.

The straight lines drawn inside the ellipse are the *axes*. The horizontal one is the *major axis*, since it is longer. The *minor axis* here is the vertical line. In many ellipses, the major axis is vertical. Each axis is a line of symmetry, a line that divides the ellipse in two halves, each of which is the mirror image of the other. The point where the axes intersect is the center of the ellipse.

If an ellipse on the coordinate plane with a horizontal major axis is centered at (h, k), its equation is

$$\frac{(x - h)^2}{a^2} + \frac{(y - k)^2}{b^2} = 1$$

If an ellipse on the coordinate plane with a vertical major axis is centered at (h, k), its equation is

$$\frac{(x - h)^2}{b^2} + \frac{(y - k)^2}{a^2} = 1$$

For all ellipses, $a^2 > b^2$. An equation where $a^2 = b^2$ would actually be that of a circle, rather than an ellipse.

The length of the minor axis of an ellipse is $2b$.

The location of the foci depends on the value of a variable c. For an ellipse with a horizontal major axis, $c^2 = a^2 - b^2$, and the coordinates of the foci are $(h \pm c, k)$.

For an ellipse with a vertical major axis, $c^2 = a^2 - b^2$, and the coordinates of the foci are $(h, k \pm c)$.

iv. The Parabola

We introduced parabolas back in Chapter 3, as the graphs of quadratic functions. As we saw there, a parabola is an umbrella-shaped curve. Unlike circles and ellipses, a parabola is not closed. In terms of conic sections, the parabola is the figure created by a plane parallel to a side of the cone, intersecting a single nappe.

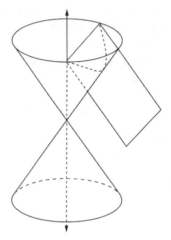

A parabola can also be defined as the set of points that are the same distance from a single focus and a line, called the *directrix*.

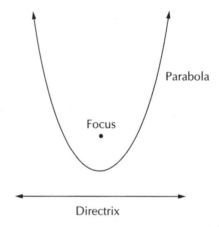

Each point on the parabola is equally distant from the closest point to it on the directrix and the focus.

Unlike the other conic sections we're studying in this chapter, the y-coordinates of a parabola are a *function* of the x-coordinates. No graph of a circle, ellipse, or hyperbola would pass the vertical line test.

Take a quadratic function $y = ax^2 + bx + c$, which is the equation of a parabola. As we saw in Chapter 3, you can rewrite that function as $y = a(x - h)^2 + k$. Given that form of the equation, you can use the values of h and k to identify many properties of a parabola.

In Chapter 3, we pointed out that the direction in which the parabola opens depends on the value of a. If a is positive, then it opens upward. If not, it opens downward. The vertex of a parabola, which is the lowest point if it opens upward or the highest point if it opens downward, has the coordinates (h, k).

The axis of symmetry is a vertical line passing through the vertex of a parabola. One side of the parabola is a mirror image of the other, reflected over the axis of symmetry. The equation of that line is $x = h$.

The focus of a parabola has the coordinates $\left(h, k + \dfrac{1}{4a}\right)$.

The directrix of the parabola has the equation $y = k - \dfrac{1}{4a}$.

v. The Hyperbola

A hyperbola is created by a plane intersecting both nappes of a double cone. The figure consists of two umbrella-shaped curves, opening in opposite directions.

Like an ellipse, a hyperbola has two foci. For any point on the hyperbola, the absolute value of the difference between the distances from each foci is a fixed number.

There are several major features of the hyperbola about which you'll need to know. A hyperbola has two axes of symmetry. Each curve is a reflection of the other. The line segment over which they are reflected is the *conjugate axis*. The line that includes the vertex of each curve is the *transverse axis*.

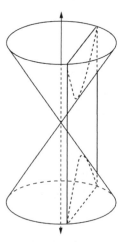

The *center* of the hyperbola is halfway between the foci. Both axes intersect at that point.

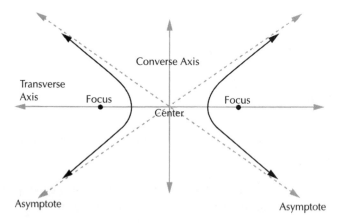

Each hyperbola has two *asymptotes*. They are diagonal lines that intersect at the center of the hyperbola. As each curve of the hyperbola opens further, it gets closer and closer to the asymptotes, but never intersects them.

Where $2a$ is the absolute value of difference of the distances between any point and the two foci, and $2b$ is the length of the converse axis, an ellipse centered at (h, k), with a horizontal transverse axis, has the equation

$$\frac{(x - h)^2}{a^2} - \frac{(y - k)^2}{b^2} = 1$$

If the transverse axis is vertical, then the equation is $\dfrac{(y - k)^2}{a^2} - \dfrac{(x - h)^2}{b^2} = 1$

The foci of an ellipse with a horizontal transverse axis are located at $(h \pm c, k)$, where $c^2 = a^2 + b^2$.

The foci of an ellipse with a vertical transverse axis are located at $(h, k \pm c)$, where $c^2 = a^2 + b^2$.

If the ellipse has a horizontal transverse axis, then the following equations apply:

Equations of asymptotes: $y = \pm \dfrac{b}{a} x$

Equation of transverse axis: $y = k$

If the ellipse has a vertical transverse axis, then the following equations apply:

Equations of asymptotes: $y = \pm \dfrac{a}{b} x$

Equation of transverse axis: $x = h$

STEPS YOU NEED TO REMEMBER

Most conic sections involve either (a) using information (either written or graphic) to derive an equation, or (b) getting information from an equation to describe a particular conic section. Below, we'll review the basic steps involved in each of those processes.

i. Using the Right Equation

The crucial first step in setting up the solution to a conic section question is identifying the right equation. We've studied four kinds of conic sections. In the cases of ellipses and hyperbolas, as we've seen, the form of an equation can vary, depending on the directions of the axes. In addition, you'll need to know the equations describing the relationships among a, b, and c in ellipses and hyperbolas. Using the wrong equation will lead you to incorrect results.

ii. Applying Features of Conic Sections to Equations

Many conic section questions will require you to plug given information into an equation. At times, you may need to solve the equation for an unknown, based on the limited information provided.

Solving an equation for an unknown value of h, k, or, where applicable, a, b, or r, requires a pair of values for x and y. The values of x and y always represent the coordinates of a point on a given conic section. If you are solving for an unknown, then the question must provide at least one point for you to use. Remember to plug in that pair of values for x and y, not h and k. Remember also that a conic section never includes the point (h, k).

iii. Analyzing Equations to Identify Features of Conic Sections

For each kind of conic section, there is a set of features you should be able to identify. For circles, there is the length of the radius, the location of the center, and the coordinates of the points. For ellipses, there are the lengths of the axes, the direction of each axis, the location of the foci, and the coordinates of the points. Parabolas and hyperbolas each have their own features too.

To determine whether a conic section includes a given pair of coordinates, all you need to do is plug the values into the right equation. If the equation is satisfied, then the figure includes that point.

Identifying other features of a conic section is a matter of reading the right numbers off of an equation. When it comes to finding the center of a figure, that's all there is to it. When it comes to identifying other features, such as the asymptotes of a hyperbola, you'll need to plug certain values into yet another equation. All of the equations you'll need to work with are found above in the **Concepts to Help You** section.

STEP-BY-STEP ILLUSTRATION OF THE FIVE MOST COMMON QUESTION TYPES

The conic sections we've discussed have many properties. The questions that follow spotlight the most common kinds of things you'll be asked to do in this area of Algebra II. That includes recognizing the properties of conic sections and their equations. It also includes deriving equations from given information, and determining certain properties of conic sections from their equations.

Question 1: Identifying a Conic Section

Which of the following conic sections does the equation

$$\frac{(x + 21)^2}{9} + \frac{y^2}{81} = 1 \text{ represent?}$$

(A) A hyperbola with a diagonal transverse axis.

(B) A hyperbola with a horizontal transverse axis.

(C) A hyperbola with a vertical transverse axis.

(D) An ellipse with a horizontal major axis.

(E) An ellipse with a vertical major axis.

The equation for an ellipse is $\frac{(x - h)^2}{a^2} + \frac{(y - k)^2}{b^2} = 1$ or $\frac{(x - h)^2}{b^2} + \frac{(y - k)^2}{a^2} = 1$. Since $\frac{(x + 21)^2}{9} + \frac{y^2}{81} = 1$ has that form, where $h = -21$ and $k = 0$, the equation is of an ellipse.

In an ellipse equation, $a^2 > b^2$, so $a^2 = 81$ and $b^2 = 9$. Thus, this equation fits the form $\frac{(x - h)^2}{b^2} + \frac{(y - k)^2}{a^2} = 1$. That is the equation of an ellipse with a vertical major axis. $\frac{(x - h)^2}{a^2} + \frac{(y - k)^2}{b^2} = 1$ is the equation of an ellipse with a horizontal major axis. **So choice (E) is the correct answer.**

The equation for a hyperbola (with a horizontal transverse axis) is
$\frac{(x-h)^2}{a^2} - \frac{(y-k)^2}{b^2} = 1$. Note the key difference between that and the
ellipse equation: one fraction is subtracted from the other in the hyperbola
equation, while they are added in the ellipse equation. They are similar in
other respects, especially in that 1 is usually the value on the right side of
each equation.

It is also worth noting that the transverse axis of a hyperbola is either
horizontal or vertical. So you could have ruled out choice (A) even if the
equation was of a hyperbola.

Question 2: Identifying Points on a Circle

A point on the circle $(x-6)^2 + (y+9)^2 = 289$ could have which
pair of coordinates?

(A) $(7, -5)$

(B) $(10, 4)$

(C) $(13, 7)$

(D) $(17, 1)$

(E) $(21, -1)$

A pair of coordinates represents values for x and y in the circle equation. To
determine whether a point at (x, y) lies on a given circle $(x-h)^2 + (y-k)^2 = r^2$,
evaluate $(x-h)^2 + (y-k)^2$. If the value equals r^2, then the circle includes the
point. Here $r^2 = 289$. Therefore, we should evaluate $(x-6)^2 + (y+9)^2$ for each
pair of values in the answer choices:

$(7, -5)$: $(x-6)^2 + (y+9)^2 = (7-6)^2 + (-5+9)^2 = 1^2 + 4^2 = 1 + 16 = 17$

$(10, 4)$: $(x-6)^2 + (y+9)^2 = (10-6)^2 + (4+9)^2 = 4^2 + 13^2 = 16 + 169 = 185$

$(13, 7)$: $(x-6)^2 + (y+9)^2 = (13-6)^2 + (7+9)^2 = 7^2 + 16^2 = 49 + 256 = 305$

$(17, 1)$: $(x-6)^2 + (y+9)^2 = (17-6)^2 + (1+9)^2 = 11^2 + 10^2 = 121 + 100 = 221$

$(21, -1)$: $(x-6)^2 + (y+9)^2 = (21-6)^2 + (-1+9)^2 = 15^2 + 8^2 = 225 + 64 = \mathbf{289}$

Since $(x-6)^2 + (y+9)^2 = 289$ when $x = 21$ and $y = -1$, the circle includes the point $(21, -1)$, and **choice (E) is the correct answer.**

Choice (A) might have been tempting because $(x-6)^2 + (y+9)^2$ has the value of the radius, 17, when $x = 7$ and $y = -5$. Remember that you need to find an expression with the value of the square of the radius. You must find values for the variables that make the given equation true.

Question 3: Properties of Ellipses

What are the coordinates of the foci of the ellipse shown below?

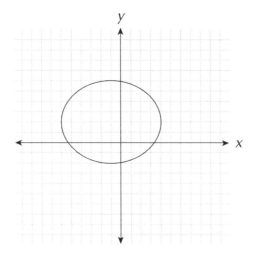

(A) $(-5, 2)$ and $(3, 2)$

(B) $(-4, 2)$ and $(2, 2)$

(C) $(-3, 2)$ and $(1, 2)$

(D) $(-2, 3)$ and $(-2, 1)$

(E) $(-2, 4)$ and $(-2, 0)$

We can find the coordinates of the foci by using the equation of the ellipse. Before we can do that, however, we'll have to piece that equation together from the information in the graph.

The center of the ellipse is at (–1, 2). The minor axis is the line segment connecting the points (–1, 6) and (–1, –2). The axis is 8 units long, so $2b = 8$, and $b = 4$. So far, then, we have the equation $\dfrac{(x-(-1))^2}{a^2} + \dfrac{(y-2)^2}{4^2} = 1$

Now, we can select a point on the ellipse, and plug its coordinates into this equation. We can then solve the equation for a^2. Let's use the coordinates (4, 2):

$$\frac{(4-(-1))^2}{a^2} + \frac{(2-2)^2}{4^2} = 1$$

$$\frac{(4+1)^2}{a^2} + \frac{(0)^2}{4^2} = 1$$

$$\frac{5^2}{a^2} + 0 = 1$$

$$a^2 = 25$$

The equation of this ellipse is $\dfrac{(x+1)^2}{25} + \dfrac{(y-2)^2}{16} = 1$

So, $c^2 = a^2 - b^2 = 25 - 16 = 9$, and $c = 3$.

The coordinates of the foci of an ellipse with a horizontal major axis are ($h \pm c$, k). Since $c = 3$, $h = -1$, and $k = 2$, the foci of this ellipse are located at (–4, 2) and (2, 2). **Choice (B) is the correct answer.**

Question 4: Properties of Parabolas

What is the equation of the directrix of the parabola $y = 2x^2 - 16x + 35$?

(A) $y = 2\frac{1}{4}$

(B) $y = 2\frac{3}{4}$

(C) $y = 2\frac{7}{8}$

(D) $y = 3\frac{1}{8}$

(E) $y = 3\frac{1}{4}$

The equation of the directrix of a parabola $y = a(x - h)^2 + k$ *is* $y = k - \dfrac{1}{4a}$.

To find the value of k, then, we must rewrite $y = 2x^2 - 16x + 35$ in the form $y = a(x - h)^2 + k$.

x^2 is multiplied by 2, so that is the value of a. Since $2(x - h)^2 + k = 2(x^2 - 2h + h^2) + k$, $2x^2 - 4hx + 2h^2 + k = 2x^2 - 16x + 35$. This means that $-4h = -16$ and $2h^2 + k = 35$.

Since $-4h = -16$, $h = 4$. Therefore, $2h^2 = 32$. Since $2h^2 + k = 35$, $k = 3$.

$y = k - \dfrac{1}{4a}$, $k = 3$, and $a = 2$. So, $y = 3 - \dfrac{1}{4(2)} = 3 - \dfrac{1}{8} = 2\dfrac{7}{8}$, and **choice (C) is the correct answer.** Choice (B) uses the value of $k - \dfrac{1}{2a}$ instead $k - \dfrac{1}{4a}$. Choice (D) uses the value $k + \dfrac{1}{4a}$.

Question 5: Properties of Hyperbolas

A hyperbola including the point (7, –14) has foci at (–2, –14) and (–2, 26). What is the equation of the hyperbola?

(A) $\dfrac{(x + 2)^2}{144} - \dfrac{(y - 6)^2}{256} = 1$

(B) $\dfrac{(x + 2)^2}{400} - \dfrac{(y - 6)^2}{144} = 1$

(C) $\dfrac{(y - 6)^2}{144} - \dfrac{(x + 2)^2}{400} = 1$

(D) $\dfrac{(y - 6)^2}{256} - \dfrac{(x + 2)^2}{144} = 1$

(E) $\dfrac{(y - 6)^2}{400} - \dfrac{(x + 2)^2}{256} = 1$

Although the question provides very little information about the hyperbola, it is enough to derive an equation. Working one step at a time, we can find the values of a and b.

Since the foci are at points (–2, –14) and (–2, 26), the hyperbola has a vertical transverse axis. The center of the ellipse is the midpoint of the line segment connecting the foci, which is (–2, 6). The foci have coordinates (h, $k \pm c$): $k + c = 26$, and $k - c = -14$. It follows that $k = 26 - c$ and $k = -14 + c$. So, $26 - c = -14 + c$, and $2c = 40$. Therefore, $c = 20$.

Next, we can use the coordinates of other points we were given, (7, –14), to find $2a$. Recall that $2a$ is the absolute value of the difference of the distances between any point on the hyperbola and the two foci. Let's go ahead and calculate the distances between the one point we have and each focus. We'll use the distance formula:

$$d = \sqrt{(x_2 - x_1)^2 + (y_2 - y_1)^2}$$

The distance between the point and the focus at (–2, –14) is

$$d_1 = \sqrt{(7 - (-2))^2 + (-14 - (-14))^2} = \sqrt{(7 + 2)^2 + (-14 + 14)^2}$$
$$\sqrt{(9)^2 + (0)^2} = \sqrt{81} = 9$$

The distance between the point and the focus at (–2, 26) is

$$d_2 = \sqrt{(7 - (-2))^2 + (-14 - 26)^2} = \sqrt{(7 + 2)^2 + (-14 - 26)^2}$$
$$\sqrt{(9)^2 + (-40)^2} = \sqrt{81 + 1600} = \sqrt{1681} = 41$$

Since $d_1 = 9$ and $d_2 = 2a = |d_1 - d_2| = |9 - 41| = |-32| = 32$ So, $a = 16$.

Now we can use the values of a and c to find the value of b. For an ellipse $\frac{(y - k)^2}{a^2} - \frac{(x - h)^2}{b^2} = 1$, $c^2 = a^2 + b^2$. So, $b^2 = c^2 - a^2$. Since $a = 16$ and $c = 20$, $b^2 = 20^2 - 16^2 = 400 - 256 = 144$. Therefore, b has a value of 12.

Now we have all of the values we need to write the equation of the hyperbola: $a = 16$, $b = 12$, $h = -2$, and $k = 6$.

$$\frac{(y-k)^2}{a^2} - \frac{(x-h)^2}{b^2} = \frac{(y-6)^2}{16^2} - \frac{(x-(-2))^2}{12^2} = \frac{(y-6)^2}{256} - \frac{(x+2)^2}{144} = 1$$

Choice (D) is the correct answer. The equation in choice (A) represents a hyperbola with a horizontal transverse axis, with the values of *a* and *b* reversed. Choice (C) uses the value of *b* instead of the value of *a* and the value of *c* instead of *b*.

CHAPTER QUIZ

1. Which of the following conic sections does the equation

   $$\frac{(x-10)^2}{6} + \frac{(y-5)^2}{6} = 24$$

   represent?

   (A) A circle

   (B) An ellipse with a horizontal major axis.

   (C) An ellipse with a vertical major axis.

   (D) A hyperbola

   (E) A parabola

2. Which of the following equations represents a hyperbola?

   (A) $\left(x + \dfrac{1}{2}\right)^2 + \left(y - \dfrac{7}{10}\right)^2 = 4$

   (B) $\dfrac{(x-3)^2}{(y+4)^2} = 4$

   (C) $y = \dfrac{5x^2}{3} - \dfrac{x}{4} + 9$

   (D) $\dfrac{(x+3)^2}{16} - \dfrac{(y+5)^2}{9} = 1$

   (E) $\dfrac{(x-1)^2}{25} + \dfrac{(y+7)^2}{100} = 1$

3. What is the radius of the circle $(x + 16)^2 + y^2 = 36$?

   (A) 4

   (B) 6

   (C) 16

   (D) 18

   (E) 36

4. The figure graphed below is a circle.

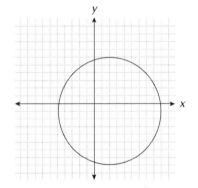

   What is the equation of the circle?

   (A) $(x-2)^2 + (y+1)^2 = 7$

   (B) $(x-2)^2 + (y-1)^2 = 14$

   (C) $(x-2)^2 + (y+1)^2 = 49$

   (D) $(x+2)^2 + (y-1)^2 = 14$

   (E) $(x+2)^2 + (y-1)^2 = 49$

5. The ellipse

$$\frac{(x - 2)^2}{100} + \frac{(y + 9)^2}{25} = 1$$

includes which of the following points?

(A) $(4, -1)$

(B) $(6, -3)$

(C) $(8, -5)$

(D) $(10, 7)$

(E) $(12, 16)$

6. Which ellipse has a minor axis with a length of 12?

(A) $\dfrac{(x + 7)^2}{4} + \dfrac{(y + 5)^2}{3} = 1$

(B) $\dfrac{(x - 9)^2}{9} + \dfrac{(y + 6)^2}{6} = 1$

(C) $\dfrac{(x - 1)^2}{36} + \dfrac{(y + 8)^2}{4} = 1$

(D) $\dfrac{(x - 5)^2}{81} + \dfrac{(y + 5)^2}{36} = 1$

(E) $\dfrac{(x - 9)^2}{256} + \dfrac{(y + 6)^2}{144} = 1$

7. What is the axis of symmetry of the parabola $y = x^2 + 14x + 40$?

(A) $x = -9$

(B) $x = -7$

(C) $x = 14$

(D) $x = 40$

(E) $x = 49$

8. What are the coordinates of the focus of the parabola $y = x^2 + 16x - 36$?

(A) $(-8, -100.25)$

(B) $(-8, -97.75)$

(C) $(-8, 100.25)$

(D) $(8, -100.25)$

(E) $(8, -97.75)$

9. What are the equations of the asymptotes of the hyperbola

$$\frac{(x - 8)^2}{64} - \frac{(y + 6)^2}{4}?$$

(A) $y = \dfrac{x}{16}$ $y = -\dfrac{x}{16}$

(B) $y = \dfrac{x}{4}$ $y = -\dfrac{x}{4}$

(C) $y = \dfrac{3x}{4}$ $y = -\dfrac{3x}{4}$

(D) $y = 4x$ and $y = -4x$

(E) $y = 16x$ and $y = -16x$

10. What is the equation of the transverse axis of the hyperbola

$$\frac{(y + 9)^2}{81} - \frac{(x + 11)^2}{49} = 1?$$

(A) $x = -11$

(B) $y = -9$

(C) $x = 7$

(D) $y = 9$

(E) $x = 11$

ANSWER EXPLANATIONS

1. A

A circle has an equation of the form $(x - h)^2 + (y - k)^2 = r^2$. Although $\dfrac{(x - 10)^2}{6} + \dfrac{(y - 5)^2}{6} = 24$ is not in that form, each of the terms on the left side has a denominator of 6. If you multiply both sides of the equation by 6, you get $(x - 10)^2 + (y - 5)^2 = 144$. If $r^2 = 144$, then $r = 12$. So, this equation represents a circle centered at $(10, 5)$ with a radius of 12.

The equation would represent an ellipse, rather than a circle, if the denominators on the left side were not equal. The major axis would be vertical if the first denominator were greater than the second. Otherwise, the major axis would be horizontal. The equation would represent a hyperbola if it had a minus sign instead of a plus sign.

2. D

The hyperbola centered at (h, k) with a horizontal transverse axis has the equation $\dfrac{(x - h)^2}{a^2} - \dfrac{(y - k)^2}{b^2} = 1$. Out of all of the equations in the answer choices, only $\dfrac{(x + 3)^2}{16} - \dfrac{(y + 5)^2}{9} = 1$ has that basic form. The equation in choice (A) represents a circle. Choice (B) does not involve a conic section; it is actually a linear equation. The equation in (C) is one of a parabola, while (E) is the equation of an ellipse.

3. B

The circle $(x - h)^2 + (y - k)^2 = r^2$ has a center at (h, k) and radius r. Since $r^2 = 36$, $r = \sqrt{36} = 6$. Choice (A) is the result of taking r^2 to equal 16 instead of 36.

4. C

The circle has a radius of 7 units and is centered at $(2, -1)$. The circle equation is $(x - h)^2 + (y - k)^2 = r^2$. Since $h = 2$, $k = -1$, and $r = 7$, the equation of this circle is $(x - 2)^2 + (y - (-1))^2 = 7^2$, or $(x - 2)^2 + (y + 1)^2 = 49$. Choice (A) uses the radius instead of its square in the equation. Choice (E) is an equation where h and k are added instead of subtracted.

5. C

Evaluate the left side of the equation for each pair of coordinates.

$(4, -1)$:

$$\frac{(x - 2)^2}{100} + \frac{(y + 9)^2}{25} = \frac{(4 - 2)^2}{100} + \frac{(-1 + 9)^2}{25} = \frac{2^2}{100} + \frac{8^2}{25} = \frac{4}{100} + \frac{64}{25} = \frac{258}{100}$$

$(6, -3)$:

$$\frac{(x - 2)^2}{100} + \frac{(y + 9)^2}{25} = \frac{(6 - 2)^2}{100} + \frac{(-3 + 9)^2}{25} = \frac{4^2}{100} + \frac{6^2}{25} = \frac{16}{100} + \frac{36}{25} = \frac{160}{100}$$

$(8, -5)$:

$$\frac{(x - 2)^2}{100} + \frac{(y + 9)^2}{25} = \frac{(8 - 2)^2}{100} + \frac{(-5 + 9)^2}{25} = \frac{6^2}{100} + \frac{4^2}{25} = \frac{36}{100} + \frac{16}{25}$$
$$= \frac{9}{25} + \frac{16}{25} = \frac{25}{25} = 1$$

$(10, 7)$:

$$\frac{(x - 2)^2}{100} + \frac{(y + 9)^2}{25} = \frac{(10 - 2)^2}{100} + \frac{(7 + 9)^2}{25} = \frac{8^2}{100} + \frac{16^2}{25} = \frac{64}{100} + \frac{256}{25} = \frac{1{,}088}{100}$$

$(12, 16)$:

$$\frac{(x - 2)^2}{100} + \frac{(y + 9)^2}{25} = \frac{(12 - 2)^2}{100} + \frac{(16 + 9)^2}{25} = \frac{10^2}{100} + \frac{25^2}{25} = \frac{100}{100} + \frac{625}{25} = \frac{650}{25}$$

$\dfrac{(x-2)^2}{100} + \dfrac{(y+9)^2}{25}$ has a value of 1 when $x = 8$ and $y = -5$. So, the ellipse

includes the point at $(8, -5)$. For all of the other pairs of values given, the expression has a value greater than 1.

6. D

The length of the minor axis is $2b$. So we are looking for an equation that uses a value of b such that $2b = 12$. If $b = 6$, then $b^2 = 36$. b^2 is always the lesser of the two denominators in the equation, so the right equation is the

one where 36 is the lesser of the two. We find that $\dfrac{(x-5)^2}{81} + \dfrac{(y+5)^2}{36} = 1$

is correct. In choice (C), 36 is the greater of the denominators, so 6 is value of a there, rather than b. In choice (E), $b = 12$, since $b^2 = 144$. However, that makes $2b$, the length of the minor axis, 24.

7. B

The axis of symmetry of the parabola $y = a(x-h)^2 + k$ is the line $x = h$. To find the axis of symmetry of $y = x^2 + 14x + 40$, we need to get the equation into that other form. Since the coefficient of the second-order term is 1, that is the value of a. Since $(x+7)^2 = x^2 + 14x + 49$, $x^2 + 14x + 40 = (x+7)^2 - 9$. Therefore, $h = -7$, and the axis of symmetry is $x = -7$. Choice (A) uses the value of k instead of h in the equation.

8. B

The focus of a parabola $y = a(x-h)^2 + k$ has coordinates $\left(h, k + \dfrac{1}{4a}\right)$. To locate the focus of the parabola $y = x^2 + 16x - 36$, rewrite it in the form $y = a(x-h)^2 + k$. Since $a = 1$, we can simply use $y = (x-h)^2 + k$.

If $-2h = 16$, then $h = -8$, and $h^2 = 64$. So, if $y = x^2 + 16x - 36 = x^2 + 16 + 64 + k$, then $k = -100$. Since $a = 1$, $k + \dfrac{1}{4a} = -100 + \dfrac{1}{4} = -97.75$, and the focus has the coordinates $(-8, -97.75)$.

Choices (A) and (D) use the value of $k - \dfrac{1}{4a}$ instead of $k + \dfrac{1}{4a}$. Choices

(D) and (E) use the value of $-h$ instead of h.

9. B

For hyperbola $\dfrac{(x-k)^2}{a^2} - \dfrac{(y-h)^2}{b^2} = 1$ (one with a horizontal transverse

axis), the linear equations of the asymptotes are $y = \pm\dfrac{b}{a}x$. Since $a^2 =$

64 and $b^2 = 4$, $a = 8$ and $b = 2$. $y = \pm\dfrac{b}{a}x = \pm\dfrac{2}{8}x = \pm\dfrac{x}{4}$ Choice (A) is

the result of using the equations $y = \pm\dfrac{b^2}{a^2}x$ instead $y = \pm\dfrac{b}{a}x$. Choice

(C) gives the result $y = \pm\dfrac{h}{a}x$. (D) gives the equations for the hyperbola

$\dfrac{(y-8)^2}{64} - \dfrac{(x+6)^2}{4}$; if the transverse axis is vertical, then the equations of

the asymptotes are given by $y = \pm\dfrac{a}{b}x$.

10. A

$\dfrac{(y+9)^2}{81} - \dfrac{(x+11)^2}{49} = 1$ is a hyperbola with a vertical transverse axis.

The equation of that axis is $x = h$. Here, $h = -11$. So the equation of the

axis is $x = -11$. Choice (B) is the equation of the transverse axis of the

$\dfrac{(x+9)^2}{81} - \dfrac{(y+11)^2}{49} = 1$. Choice (D) uses the equation $y = a$ instead of $x = h$.

Exponents and Logarithms

WHAT ARE EXPONENTS AND LOGARITHMS?

The main focus of this chapter is logarithms, but they are best understood in terms of *exponential functions*. An exponential function is one of the form $y = b^x$, where b is a real number. In that equation, $y = b^x$, b is the base, and x is the exponent. The exponent raises the base to a certain power. A logarithmic function is very much related. A logarithm gives the value of the exponent x in terms of the number y and the base b.

If $y = b^x$, then $\log_b y = x$, and vice versa.

CONCEPTS TO HELP YOU

Although you may be long familiar with exponents, logarithms don't often come up before Algebra II. We'll make the key logarithmic concepts clear in terms of exponents.

i. Logarithms

As we've said, a logarithm is basically an exponent. It is the number that raises a base to a certain power. The operator *log* is commonly used to indicate a logarithm. Following the operator, you will often find a number or variable in subscript. That is the base of the logarithm. After the equal sign, you will see another full-size number or variable. That is the power to which the base is raised.

So, the logarithm $\log_3 9$ has a value of 2 because you get 9 by raising 3 to the second power. Here, we would say that, the base 3 logarithm of 9 is 2.

Not all logarithms are integers. Many logarithms are irrational numbers; they appear as nonrepeating, nonterminating decimals.

For instance, $\log_3 10$ has an approximate value of 2.0959. You can verify that by using your scientific calculator to find that $3^{2.0959} \approx 10$.

The integer part of a logarithm is called the *mantissa*. The decimal part is called the *characteristic*.

One of the most frequently used logarithms is the base 10 logarithm. In fact, logarithms such as $\log_{10} 20$ and $\log_{10} 35$ are called *common logarithms*. They are used so commonly that the base is often left out. If you see a logarithm with no base, then you may take it to be a base 10 or common logarithm. So $\log_{10} 20$ can be written as log 20, and $\log_{10} 35$ can be written as log 35.

Many scientific calculators and logarithmic tables are set to give the values of base 10 logarithms. Most of the work we'll do in the rest of this chapter involves common logarithms, so you'll see many logarithmic functions without bases.

ii. Antilogarithms

As the name suggests, the antilogarithm is the inverse of the logarithmic function. It usually involves the operator *antilog*. If log $x = 10$, then antilog $10 = x$. We'll deal with common antilogarithms primarily (i.e., base 10 antilogarithms). Unless you happen to recognize an antilogarithm as an exponent, you'll likely need the assistance of a scientific calculator or logarithmic table to find the antilogarithm of a given value.

iii. The Power Property of Logarithms

Since $10^2 = 100$, $\log_{10} 100$ (log 100) has a value of 2. Since $10^6 = 1,000,000$, log $1,000,000 = 6$. Note that 6 is 2 multiplied by 3. Thus, log $1,000,000 = 3$ log 100 (log 100 multiplied by 3.)

Now, note that $1,000,000 = 100^3$. That this relation holds along with log $1,000,000 = 3$ log 100 reflects an important property of logarithms. The *power property of logarithms* holds that log $x^n = n$log x. Whenever you have the logarithm of a power, you can move the exponent to the outside of the logarithm, such that it multiplies the logarithm of the base. Thus, log $16 = $ log $4^2 = 2$ log 4, and log $125 = $ log $5^3 = 3$ log 5.

iv. The Product and Quotient Properties of Logarithms

Since $10^5 = 100,000$, log $100,000 = 5$. We saw above that log $100 = 2$, since $10^2 = 100$. We can also determine that log $1,000 = 3$. We can make another important observation here: the product of 100 and 1,000 is 100,000, and the sum of the logarithms of 100 and 1,000 is the logarithm of 100,000 ($2 + 3 = 5$).

In general, if a number is the product of two factors, then the logarithm of the number is the sum of the logarithms of the factors:

log mn = log m + log n

So, log 30 = log 5 + log 6, since 5 and 6 have a product of 30. This is the *product property of logarithms*.

A related property, the *quotient property of logarithms*, holds that $\log \frac{m}{n}$ = log m – log n. By this property, log 18 = log 36 – log 2, since $36 \div 2 = 18$. You can verify that by finding the values of the logarithms:

log $36 \approx 1.5563$

log $2 \approx 0.3010$

log $18 \approx 1.2553$

log 36 – log $2 \approx 1.5563 - 0.3010 \approx 1.2553$

antilog $1.2553 \approx 18$

THE POWER PROPERTY AND THE QUOTIENT PROPERTY

You can think about the power property in terms of the product property. Take log 49, which is the same as log 7^2, or 2 log 7. By the power property, log 49 = log 7 + log 7, because log 49 = log (7)(7). Since log 7 + log 7 = 2 log 7, we can see that we can get the same result with the power property as we did with the product property.

STEPS YOU NEED TO REMEMBER

To set up the proper solution to many exponent and logarithm questions, you'll need to apply the properties of logarithms carefully. In particular, the use of the power property and the antilogarithm function are steps you'll need to apply correctly.

i. Evaluating Logarithms

Evaluating a logarithm is usually a matter of solving an equation. That's because it usually helps to put a logarithm in terms of an exponential equation. If $y = \log_b n$ and you need to find the value of y given b and n, then you should plug your values into the equation $b^y = n$. It's easier to see how the unknown relates to the other values in an exponential equation. If you know what power you need to raise b to in order to get n, then you're done. If not, then you can make your logarithm a common logarithm.

If $b^y = n$, then $\log b^y = \log n$. Therefore, you can find the value of y in $\log_b n = y$ by solving $\log b^y = \log n$. We'll show you how to do that soon.

ii. Using Logarithms to Evaluate Exponents

You can use logarithms to find the value of a power b^p. Logarithms can be helpful here when either the base b or the exponent p is a noninteger. You may be able to evaluate 4^3 in your head, but $4^{3.2}$ is much tougher to deal with!

The power property of logarithms comes in handy here. Start with the logarithm $\log 4^{3.2}$. No calculator or table will give you the logarithm of that expression directly, but you work with $3.2 \log 4$ instead, thanks to the power property. Multiply the logarithm of 4 by 3.2 to get the value of $\log 4^{3.2}$. Finally, get the antilogarithm of that value to find $4^{3.2}$. Your result will be an approximate one, since you will be dealing with irrational numbers.

iii. Using Logarithms to Solve Equations

Suppose you are given the equation $x^{2.5} = 40$. It's unlikely that you can figure out in your head what number raised to the power of 2.5 gets you 40. You can set up a logarithmic equation to find that unknown, though. If $x^{2.5} = 40$, then $\log x^{2.5} = \log 40$. To find the value of x, your first goal should be to find the value of $\log x$ by getting that expression alone on one side of the equation. Since $\log x^{2.5} = \log 40$, $2.5 \log x = \log 40$, and $\log x = \dfrac{\log 40}{2.5}$. Once you find the value of $\dfrac{\log 40}{2.5}$, you can take the antilogarithm to find the value of x.

If you are instead working with an exponential equation with an unknown exponent, such as $2.5^x = 40$, the strategy is very much the same. Take the logarithm of each side of the equation to get $\log 2.5^x = \log 40$. By the power property, it follows that $x \log 2.5 = \log 40$.

Divide both sides by $\log 2.5$ to get $x = \dfrac{\log 40}{\log 2.5}$. Since you have x on the left side of the equation, you're done as soon as you evaluate the right side of the equation. These are the steps you would follow to solve the equation $\log b^y = \log n$ we started with in section **i** above.

STEP-BY-STEP ILLUSTRATION OF THE FIVE MOST COMMON QUESTION TYPES

Now it's time to tackle a variety of questions involving exponents and logarithms. Some questions require you to use logarithms to find precise or approximate values. Other questions require you to apply your knowledge of the properties of logarithms, without having you actually evaluate them.

Question 1: Evaluating Logarithms

What is the value of $\log_3 243$?

(A) 5

(B) 9

(C) 24

(D) 45

(E) 81

We can evaluate this logarithm by putting it in terms of an exponential equation. If $\log_b x = y$, then $x = b^y$. Since $\log_3 243 = y$, $3^y = 243$.

$$\log 3^y = \log 243$$

$$y \log 3 = \log 243$$

$$y = \frac{\log 243}{\log 3} = 5$$

So, y is the power to which 3 is raised to get 243. Since $3^5 = 243$, $y = 5$, **choice (A) is the correct answer.**

Question 2: Finding the Base of a Logarithm

If $log_x 14 = 0.5$, then $x =$

(A) 7

(B) 28

(C) 56

(D) 98

(E) 196

This logarithmic equation states that 14 is the value of an unknown number raised to the power of 0.5. The value of a power with the exponent 0.5 is the square root of the base. Therefore, 14 is the square root of x:

$$log_x 14 = 0.5 \rightarrow x^{0.5} = 14 \rightarrow \sqrt{x} = 14$$

Since 14 is the square root of x, the value of x must be the square of 14.

$$\sqrt{x} = 14 \rightarrow \left(\sqrt{x}\right)^2 = 14^2 = 196.$$ **Choice (E) is the correct answer.**

Question 3: Using Logarithms to Evaluate Powers

$\log 2 \approx 0.3010$. What is the approximate value of $2^{4.6}$?

(A) 1.38

(B) 8.36

(C) 17.51

(D) 21.16

(E) 24.25

This question is presented so that a calculator need not be used. With a scientific calculator, we could just find the value of $2^{4.6}$ and take its logarithm. As long as we have the value of $\log 2$, however, we can find the value of $\log 2^{4.6}$ without a calculator.

We can evaluate this power by using the power property of logarithms. We will find the value of a certain logarithm, and then take the antilogarithm of that value.

First, we'll take the logarithm of our power, $\log 2^{4.6}$. According to the power property, $\log 2^{4.6} = 4.6 \log 2$. The value of $\log 2$ is approximately 0.3010, so $4.6 \log 2 \approx 4.6(0.3010) \approx 1.3847$.

Note that this is not the answer to the question, even though it appears in choice (A). Rather, it is the next to last step. What we have here is a logarithm; we are still working with the logarithm of $2^{4.6}$. To get the actual value of $2^{4.6}$, we need to find the antilogarithm:

antilog $1.3847 \approx 24.2515$

So, choice (E) is the correct answer.

Choice (D) is the value of 4.6^2, rather than $2^{4.6}$. Remember that you can bring the exponent, not the base, to the outside of the logarithm.

Question 4: Using Logarithms to Solve an Exponential Equation

If $81^x = 19,683$, then what is the value of x?

(A) 1.25

(B) 1.75

(C) 2.25

(D) 2.75

(E) 3.25

To solve this equation, we can use the logarithm of each side. Since $81^x = 19,683$, log 81^x = log 19,683. According to the power property, log $81^x = x$ log 81.

So, x log 81 = log 19,683, and $x = \dfrac{\log 19,683}{\log 81}$

To carry out the last step, you will use the aid of a scientific calculator or table of logarithmic values.

$$\frac{\log 19,683}{\log 81} \approx \frac{4.2941}{1.9085} = 2.25$$

Thus, 81 raised to the power of 2.25 has a value of 19,683. **Choice (C) is the correct answer.**

Question 5: Using the Product Property of Logarithms

Which of the following expressions has the same value as log 144?

(A) log 10 + log 14.4

(B) $(\log 12)^2$

(C) log 72 + log 72

(D) $(\log 72)^2$

(E) log 100 + log 44

To find equivalent logarithmic expressions, you'll need to use the properties we've reviewed to rewrite the ones provided. The product property of logarithms holds that $\log ab = \log a + \log b$. So, the sum of two logarithms equals log 144 if the numbers in each have a product of 144. Since 10 x 14.4 = 144, $\log 10 + \log 14.4 = \log (10)(14.4) = \log 144$, so **choice (A) is the correct answer.**

Choice (B) is equivalent to (log 12)(log 12). According the power property, that is equivalent to $\log 12^{\log 12}$. Log 144 equals $\log 12^2$, which has a different value. The difference hinges on the parentheses in $(\log 12)^2$, which make the exponent operate on the whole logarithm.

Choice (C) is actually equivalent to $\log 72^2$. You might have picked it as a result of thinking that $\log a + \log b = \log a + b$ instead of $\log a + \log b = \log ab$. The same incorrect formula might be used to come to choice (E).

Choice (D) is equivalent to 2 log 72, which has a value different from log 2(72), or log 144.

Chapter Quiz

1. Which of the following logarithms has a value of 4?

   (A) $\log_2 2$

   (B) $\log_4 1{,}024$

   (C) $\log_5 625$

   (D) $\log_{81} 3$

   (E) $\log_{125} 5$

2. If $\log_2 x = 8$, then what is the value of x?

   (A) $\dfrac{1}{3}$

   (B) 4

   (C) 64

   (D) 256

   (E) 512

3. If $2.25^x = 3.375$, what is the value of x?

   (A) 1.25

   (B) 1.5

   (C) 1.75

   (D) 2.25

   (E) 2.5

4. Which expression has the same value as $6.5^{1.8}$?

   (A) antilog 1.8log 6.5

   (B) antilog 6.5log 1.8

   (C) antilog (log 6.5)$^{1.8}$

   (D) (antilog 1.8) (log 6.5)

   (E) (antilog 6.5) (log 1.8)

5. If $2\log_4 8 = \log_8 x$, then what is the value of x?

   (A) 8

   (B) 64

   (C) 128

   (D) 256

   (E) 512

6. If $\log 81 = x \log 3$, then what is the value of x?

   ____

7. If $\log 512 = 3 \log x$, then what is the value of x?

   ____

8. Which of the following equals log 20 + log 5?

    (A) 2 log 2

    (B) 2 log 5

    (C) 2 log 10

    (D) 2 log 25

    (E) 2 log 100

9. Which of the following is the equivalent of log 5?

    (A) log 4 – log 20

    (B) log 17 – log 12

    (C) log 25 – log 2

    (D) log 30 – log 6

    (E) log 32 – log 5

10. Which of the following equals 3 log 6?

    (A) log 18

    (B) log 72 + log 3

    (C) log 243 + log 3

    (D) 2 log 9

    (E) 6 log 2

ANSWER EXPLANATIONS

1. C

If $\log_5 625 = x$, then $5^x = 625$. Since $5^4 = 625$, $\log_5 625 = 4$.

Choice (A) has a value of 1, since 2 raised to the first power equals itself. Choice (B) has a value of 5; $\log_4 64$, rather than $\log_4 256$, has a value of 4. The value of choice (D) is $\frac{1}{4}$; while 81 is 3 to the fourth power, 3 is the fourth root of 81. Choice (E) has a value of $\frac{1}{3}$.

2. D

If $\log_2 x = 8$, $x = 2^8 = 256$.

Choice (A) is value of $\log_8 2$, since 2 is the cube root of 8. Choice (C) is the solution of $\log_8 x = 2$, since $8^2 = 64$.

3. B

If $2.25^x = 3.375$, then $\log 2.25^x = \log 3.375$, and $x \log 2.25 = \log 3.375$.

$$x = \frac{\log 3.375}{\log 2.25} = 1.5$$

4. A

Since the logarithm and the antilogarithm are inverse functions, $6.5^{1.8} =$ antilog log $6.5^{1.8}$ (the antilogarithm of a logarithm of a number is just that number). Now, by the power property, $\log 6.5^{1.8} = 1.8 \log 6.5$. Therefore, antilog log $6.5^{1.8} =$ antilog $1.8 \log 6.5$.

Choice (B) uses the power property incorrectly by bringing 6.5 instead of 1.8 to the outside. Choice (D) involves an incorrect use of parentheses. Separating 1.8 and log 6.5 with parentheses causes the antilog function to operate on just 1.8, and not on 1.8log 6.5.

5. E

$2\log_4 8 = \log_4 8^2 = \log_4 64$. Since $4^3 = 64$, $\log_4 64 = 3$. So, $\log_8 x = 3$, and $x = 8^3 = 512$.

Choice (B), 64, might be the result of taking $2\log_4 8$ to equal $\log_4 16$ instead of $\log_4 64$. That would give $\log_8 x$ a value of 2 instead of 3.

6. 4

$x \log 3 = \log 3^x$. So, $\log 81 = \log 3^x$. Since $81 = 3^4$, $x = 4$.

7. 8

By the power property, $3 \log x = \log x^3$. So $\log 512 = \log x^3$, and $512 = x^3$. Since $8^3 = 512$, $x = 8$.

8. C

By the product property, $\log 20 + \log 5 = \log (20)(5) = \log 100$. By the power property, $\log 100 = \log 10^2 = 2 \log 10$.

Choice (A) is the value of $\log 2^2$, or $\log 4$, which is what you would get if you took $\log 20 + \log 5$ to equal $\log (20 \div 5)$ instead of $\log (20)(5)$. Choice (D) is the value of $\log (20 + 5)$; remember that adding logarithms requires you to multiply the numbers.

9. D

The quotient property holds that $\log \frac{m}{n} = \log m - \log n$. Therefore, the correct answer choice has numbers with a quotient of 5. The only such pair among the choices is that of 30 and 6.

Choice (B) uses numbers with a difference of 5, but that expression actually has a value of $\log \frac{17}{12}$. You might have picked choice (C) if you took it that $\log 25 - \log 2 = \log 25^{\frac{1}{2}}$.

10. B

By the power property, $3 \log 6 = \log 6^3 = \log 216$. By the product property, $\log 216 = \log (72)(3) = \log 72 + \log 3$.

Choice (C), $\log 243 + \log 3$, has the value of $\log 729$, which is $\log 3^6$. Choice (D) has the value of $\log 81$, and (E) has the value of $\log 36$.

Trigonometric Functions

WHAT ARE TRIGONOMETRIC FUNCTIONS?

Trigonometry is the study of the relationships between angle measures and side lengths of triangles. The fundamental principle of trigonometry is that the ratios of the side lengths of a right triangle depend on its angle measures. *Trigonometric functions* relate angle measures to such ratios. Right triangles are a good basis for understanding trigonometric functions, but those functions can be applied to much more than right triangles.

CONCEPTS TO HELP YOU

Here we will introduce several trigonometric functions, and show how you can use them to find missing information about triangles, including side lengths and angle measures.

i. Right Triangles and Ratios

Look at the right triangle *ABC* below.

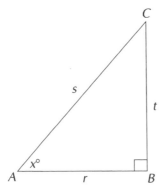

Angle A has a measure of $x°$. Side $\overline{AB} = r$, $\overline{AC} = s$, and $\overline{BC} = t$. We can find a number of ratios of side lengths here, including $\frac{r}{s}$, $\frac{s}{t}$, $\frac{t}{r}$, and $\frac{s}{r}$. Such ratios will be found in any right triangle with an angle of the same degree measure x.

ii. Sine, Cosine, and Tangent

In triangle ABC above, the ratio $\frac{t}{s}$, which is the ratio of the length of the side opposite A to the hypotenuse, is the *sine* of A. It can be expressed as a fraction or a decimal, and it must have a value greater than zero and less than one (because the hypotenuse must be longer than either of the other sides).

So, the sine function, $y = \sin x$ or $f(x) = \sin x$, has an angle measure for an argument, and pairs it with a number between zero and one.

The *cosine* of A is the ratio of the side adjacent to it, r, and the hypotenuse. So, $\frac{r}{s}$ is the cosine of A. Like the sine ratio, the range of the cosine function is the set of real numbers greater than zero and less than one.

The *tangent* of A is the ratio of the opposite side to the adjacent side. Thus, $\frac{t}{r}$ is the tangent of angle A. Unlike the sine and cosine ratios, the value of the tangent ratio can be any positive real number.

REMEMBERING THE TRIGONOMETRIC FUNCTIONS

There is a very common device that can help you recall the ratio involved in each of the three trigonometric functions.

Remember the term *SOH–CAH–TOA* (pronounced *so–kah–toe–ah*).

It stands for: sine, opposite, hypotenuse; cosine, adjacent, hypotenuse; tangent, opposite, adjacent.

If you know the lengths of the appropriate pair of sides, then you can use their ratio to find the sine, cosine, or tangent of a right triangle angle. If you can't get that information, but you have the degree measure of the angle in question, then you can find the function with the help of a scientific calculator or trigonometric function table.

OTHER TRIGONOMETRIC FUNCTIONS

In addition to the sine, cosine, and tangent, there are three other common trigonometric functions.

The *secant* of a right triangle angle is the ratio of the hypotenuse to the adjacent side. It is the reciprocal or multiplicative inverse of the cosine ratio.

The *cosecant* is the ratio of the hypotenuse to the opposite side. It is the reciprocal of the sine ratio.

The *cotangent* is the ratio of the adjacent side to the opposite side. It is the reciprocal of the tangent ratio.

iii. Inverse Trigonometric Functions

If you have the sine, cosine, or tangent of a given angle, you can find the measure of the angle with an *inverse trigonometric function*.

If x is the sine of y, then y is the *Arcsine* of x (provided that the domain of the sine function is kept to angles of 90° or less). This inverse function can be written as $y = \arcsin x$, or $y = \sin^{-1} x$.

If x is the cosine of y, then y is the *Arccosine* of x (provided that the domain of the cosine function is kept to angles of 180° or less). This inverse function can be written as $y = \arccos x$, or $y = \cos^{-1} x$.

If x is the tangent of y, then y is the *Arctangent* of x (provided that the domain of the tangent function is kept to angles of 90° or less). This inverse function can be written as $y = \arctan x$, or $y = \tan^{-1} x$.

Using these inverse functions requires the use of a scientific calculator or table of trigonometric functions.

iv. The Law of Sines and the Law of Cosines

You can use the trigonometric functions of an angle to find the measure of an angle or the length of a side of a right triangle in very few steps. However, applying trigonometric functions in non–right triangles is not as straightforward. In a non–right triangle, a trigonometric function of an angle is not the ratio of two sides. To apply trigonometric functions in such triangles, you can use the law of sines or the law of cosines.

The *law of sines* holds that the ratio of the sine of an angle to the length of its opposite side is the same for all three angles of a triangle. So, for triangle ABC, where a is the length of the opposite side of angle A, b is the length of the opposite side of angle B, and c is the length of the opposite side of angle C,

$$\frac{\sin A}{a} = \frac{\sin B}{b} = \frac{\sin C}{c}$$

You can apply the law of sines to a given triangle as long as you know, or can figure out, one of the following:

- The lengths of two sides and the measure of one of the two opposite angles
- The measures of two angles and the length of any side

If you don't have enough information to apply the law of sines, you may still be able to apply the law of cosines. According to that law, for the triangle described above, the following relationships hold:

$$a^2 = b^2 + c^2 - 2bc \cos A$$

$$b^2 = a^2 + c^2 - 2ac \cos B$$

$$c^2 = a^2 + b^2 - 2ab \cos C$$

You can apply the law of cosines to a given triangle as long as you know, or can figure out, one of the following:

- The lengths of all three sides
- The lengths of two sides and the angle where they intersect

STEPS YOU NEED TO REMEMBER

The methods we'll describe in this section will enable you to use given information about a triangle to find unknown measurements and values. In each step of the solution to a trigonometry question, you'll need to assess carefully how you can use the information you have to get the information you need.

i. Working with the Sides of Right Triangles

If you are given side measurements of a right triangle, you must be careful about using the right pair to find a particular trigonometric function. If you don't have angle measurements to work with (other than the right angle), then you need two side lengths to find a trigonometric function. Of course, once you have two side lengths, you can use the Pythagorean Theorem to find the third one. You will then be in a position to determine any of the trigonometric functions.

ii. Using the Trigonometric Functions

If you have the value of a trigonometric function of an angle, you can use it to do either of the following things:

- Find the measure of the angle. You can use the angle's inverse trigonometric function to find its degree measure.
- Find the ratio of two side lengths (in a right triangle). This is more feasible when you set the function in the form of a fraction, rather than a nonterminating, nonrepeating decimal. You can then use the Pythagorean theorem to find the ratio of either of those sides to the third side. If one of the side lengths is given, then you can use the ratios to find the other two side lengths. Also, once you have the proportions of the right triangle, you can determine the other trigonometric functions.

iii. Using the Law of Sines and the Law of Cosines

Using these laws is a matter of plugging in the information you have. Before you select a law to work with, however, you should take stock of the information you have. When you can use the law of sines or the law of cosines depends upon the conditions summarized in section **iv** of the **Concepts to Help You** section.

You can use the law of sines to solve an equation for a side length or the sine of an angle. If you solve for a sine, use the Arcsine function to get the angle measure.

You can use the law of cosines to find the cosine of an angle. You could then use the Arccosine function to find the angle measure. If you use an equation based on the law of cosines to find a side length, remember that you will first get to an equation with the square of the length in question. You have to find the square root of that number to get the side length.

STEP-BY-STEP ILLUSTRATION OF THE FIVE MOST COMMON QUESTION TYPES

Now it's time to walk through the solutions to a number of typical questions. As you'll see here, the more basic trigonometry questions tend to involve right triangles. Their solutions may still involve multiple steps though. The more difficult questions tend to involve non–right triangles, and require you to use one of the trigonometric laws we've discussed.

Question 1: Evaluating Trigonometric Functions

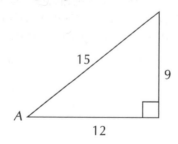

What is the sine of A in the above right triangle?

(A) $\dfrac{3}{5}$

(B) $\dfrac{3}{4}$

(C) $\dfrac{4}{5}$

(D) $\dfrac{4}{3}$

(E) $\dfrac{5}{3}$

The sine of the angle of a right triangle is the ratio of the length of the opposite side to the length of the hypotenuse. Since the length of the side opposite A is 9, and the hypotenuse has length 15, $\sin A = \dfrac{9}{15} = \dfrac{3}{5}$. **Choice (A) is the correct answer.**

Choice (B) is the tangent of A, and (C) is the cosine of that angle. (D) is the cotangent of A, and (E) is the cosecant.

Question 2: Using One Trigonometric Function to Evaluate Another

The sine of P is $\dfrac{15}{17}$. What is the value of $\tan Q$?

(A) $\dfrac{8}{17}$

(B) $\dfrac{8}{15}$

(C) $\dfrac{17}{15}$

(D) $\dfrac{15}{8}$

(E) $\dfrac{17}{8}$

The tangent of Q is the ratio of the length of the opposite side, PR, to the length of the adjacent side, QR. Since we are not told the lengths of either of those sides, we'll have to use the bit of information provided to figure them out.

The sine of an angle in a right triangle is the ratio of the opposite side length to the hypotenuse length. Since the hypotenuse has a length of 51, we can say that $\frac{15}{17} = \frac{x}{51}$, where x is the length of the opposite side. Solving that proportion gets a value of 45 for x. Now that we have the lengths of two sides of the right triangle, 51 and 45, we can use the Pythagorean theorem to find the length of PR:

$$\overline{PR} = \sqrt{51^2 - 45^2} = \sqrt{2,601 - 2,025} = \sqrt{576} = 24$$

So, we now have the lengths of all three sides:

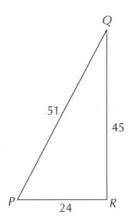

Since $\overline{PR} = 24$ and $\overline{QR} = 45$, the ratio of $\overline{PR}$ to $\overline{QR}$ is $\frac{8}{15}$. **Choice (B) is the correct answer.**

Choice (A), $\frac{8}{17}$, is actually the sine of Q. Choice (C), $\frac{17}{15}$, is the cosecant of P. Be careful to use the right function with the correct angle.

Question 3: Using Inverse Functions to Find Angle Measures

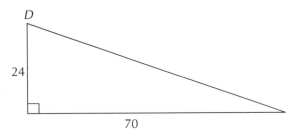

In the above right triangle, what is the approximate measure of angle *D*?

(A) 67.8°

(B) 69.9°

(C) 71.1°

(D) 72.1°

(E) 73.2°

Finding the measure of an angle of a right triangle based on the ratio of two sides requires the use of an inverse trigonometric function. We first need to find the value of one of the trigonometric functions for this figure. Since we have the lengths of the sides opposite of and adjacent to angle *D*, we can find the tangent quickly. Finding the cosine or sine of *D* would involve finding the length of the hypotenuse first.

The tangent of *D* is the ratio of the opposite side length, 70, to the adjacent side length, 24. Use your calculator to divide:

$$\frac{70}{24} \approx 2.916....$$

Applying the inverse tangent function on a scientific calculator results in an approximate value of 71.075, which can be rounded to 71.1. **Choice (C) is the correct answer.** You could also get this approximate value by finding the angle measure corresponding to the tangent 2.916 on a table of trigonometric values.

Choice (B), 69.9°, is actually the inverse cosine function of $\frac{24}{70}$. Choice (D), 72.1, is the inverse tangent function of $\frac{24}{74}$ (74 happens to be the length of the hypotenuse).

Question 4: Finding the Side Lengths of a Non–Right Triangle

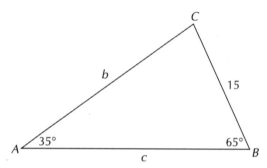

What is the approximate length of c in triangle *ABC?*

(A) 16.30

(B) 19.81

(C) 22.43

(D) 25.75

(E) 26.56

Since this is not a right triangle, we cannot apply trigonometric functions as ratios of the side lengths. However, we know the measures of two angles and the length of one side. Therefore, we can use the law of sines to find the side length in question. Here, we can use the equation $\frac{\sin A}{a} = \frac{\sin C}{c}$.

Angle *C* has a measure of 80° $(180° - (35° + 65°))$. So, $\frac{\sin 35°}{15} = \frac{\sin 80°}{c}$.

Cross–multiplying and dividing gets us the equation $c = \frac{15\sin 80°}{\sin 35°}$. The next step, evaluating the right side of the equation, requires a scientific calculator or table of trigonometric values:

$$c = \frac{15\sin 80°}{\sin 35°} \approx \frac{15(0.9848)}{0.5736} \approx \frac{14.7721}{0.5736} \approx 25.75$$

Choice (D) is the correct answer.

Choice (A), 16.30, is the value of $\dfrac{15\sin 80°}{\sin 65°}$. Remember that you need to match each angle measure with the length of the *opposite side*. Choice (E), 26.56, is the value of $c = \dfrac{15}{(\sin 80°)(\sin 35°)}$. Be careful when performing operations to solve proportions.

Question 5: Finding the Measure of an Unknown Angle in a Non–Right Triangle

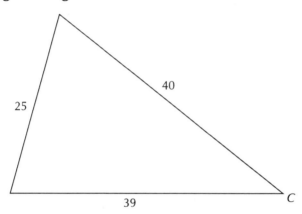

What is the approximate measure of angle *C* in the above triangle?

(A) 33.60°

(B) 34.98°

(C) 36.87°

(D) 37.55°

(E) 38.43°

Here, we are given the lengths of all three sides, and none of the angle measures. We don't have the information needed to use the law of sines, but we can use the law of cosines here:

$$c^2 = a^2 + b^2 - 2ab \cos C$$

Since c is the side opposite angle C, $c = 25$. Since angles A and B are not labeled, let's just say that $a = 39$ and $b = 40$. Now let's plug those values into the formula:

$$25^2 = 39^2 + 40^2 - 2(39)(40)\cos C$$
$$625 = 1,521 + 1,600 - 3,120 \cos C$$
$$625 = 3,121 - 3,120 \cos C$$

Next, we must solve this equation for $\cos C$:

$$625 = 3,121 - 3,120 \cos C$$
$$3,120 \cos C = 3,121 - 625$$
$$3,120 \cos C = 2,496$$
$$\cos C = \frac{2,496}{3,120}$$
$$\cos C \approx 0.8$$

Finally, we can use a calculator or table to find $\cos^{-1} 0.8$. $c \approx 36.87°$. **Choice (C) is the correct answer.**

Choice (A), 33.60°, is the result of adding 625 to 3,121 and dividing 3,120 by the sum.

CHAPTER QUIZ

1. The right triangle below has labeled side lengths.

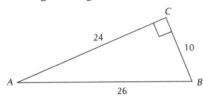

Which of the following has a value of $\frac{5}{13}$?

(A) cos *A*

(B) cos *B*

(C) sin *B*

(D) tan *A*

(E) tan *B*

2. The right triangle below has labeled side lengths.

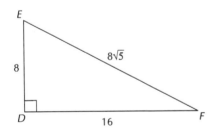

What is the value of tan *F*?

(A) $\frac{1}{2}$

(B) $\frac{1}{\sqrt{5}}$

(C) $\frac{\sqrt{5}}{2}$

(D) 2

(E) $\sqrt{5}$

3. Angle *M* in a right triangle *MNO* has a cosine of $\frac{20}{29}$ and a tangent of $\frac{21}{20}$. What is the sine of *M*?

———

4. Angle *K* in a right triangle has a sine of 0.8. What is the cosine of *K*?

(A) 0.2

(B) 0.4

(C) 0.6

(D) 1.33

(E) 1.25

5. In the right triangle below, what is the approximate measure of angle *N*?

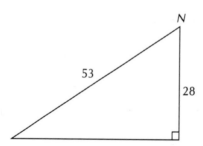

(A) 27.8°

(B) 31.9°

(C) 49.4°

(D) 58.1°

(E) 62.2°

6. In the right triangle below, what is the approximate measure of angle *R*?

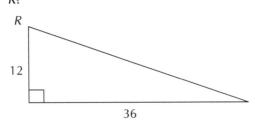

(A) 66.5°

(B) 67.3°

(C) 68.9°

(D) 70.5°

(E) 71.6°

7. A triangle with no right angles has an angle measuring 32°. The opposite side has length 12. If another side has length 18, then what is the approximate length of the third side?

(A) 22.5

(B) 23.7

(C) 24.2

(D) 25.6

(E) 27.1

8. In triangle *ABC*, *A* = 45°, *B* = 72°, and the side opposite angle *C* has length of 24. What is the length of the side opposite *A*?

(A) 19.0

(B) 24.7

(C) 25.6

(D) 27.3

(E) 32.9

9. The triangle below has sides of length $\sqrt{27}$, $\sqrt{3}$, and x.

What is the value of x?

(A) $\sqrt{15}$

(B) $\sqrt{18}$

(C) $\sqrt{21}$

(D) $\sqrt{24}$

(E) $\sqrt{30}$

10. The triangle below has side lengths 43, 61, and 68.

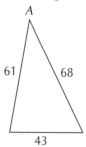

What is the approximate measure of angle *A*?

(A) 33.26°

(B) 34.66°

(C) 35.53°

(D) 37.13°

(E) 38.46°

ANSWER EXPLANATIONS

1. B

To find out which function has a value of $\frac{5}{13}$, evaluate each one:

$$\cos A = \frac{24}{26} = \frac{12}{13}$$

$$\cos B = \frac{10}{26} = \frac{5}{13}$$

$$\sin B = \frac{24}{26} = \frac{12}{13}$$

$$\tan A = \frac{10}{24} = \frac{5}{12}$$

$$\tan B = \frac{24}{10} = \frac{12}{5}$$

So the cosine of B has a value of $\frac{5}{13}$.

2. A

The tangent of angle F is the ratio of the opposite side length to the adjacent side length: $\frac{8}{16} = \frac{1}{2}$. Choice (B) is the sine of F, while (D) is the tangent of E. Choice (C) is the secant of F, and (E) is the cosecant of F.

3. $\frac{21}{29}$

The cosine of M is the ratio 20 to 29, the ratio of the adjacent side to the hypotenuse. The tangent is the ratio 21 to 20, the ratio of the opposite side to the adjacent side. So, the ratio of the opposite side to the hypotenuse is 21 to 29. Therefore, the sine of M is $\frac{21}{29}$.

4. C

If you have a scientific calculator, you can find arcsin 0.8, and find the cosine of that angle measure. If you don't have a calculator, note that since $0.8 = \frac{4}{5}$, the ratio of the opposite side to the hypotenuse is 4 to 5. Using the Pythagorean theorem, find that $5^2 - 4^2 = 3^2$. So, the ratio of the adjacent side length to the opposite side length is 3 to 4. The ratio of the adjacent side length to the hypotenuse length is 3 to 5. That makes the cosine $\frac{3}{5}$, or 0.6.

Choice (D) is the tangent of K, and (E) is the cosecant.

5. D

Here you are given the lengths of the hypotenuse and the side adjacent to the angle. Since the cosine of a right triangle angle is the ratio of the adjacent side length to the hypotenuse, the cosine of N is $\frac{28}{53}$. So, $\cos N \approx$ 0.5283, and angle $N \approx 58.1°$.

Choice (B) is the result of using $\frac{28}{53}$ as the sine of N, while (E) is the result of using $\frac{53}{28}$ as the tangent of N.

6. E

Since the side opposite R has length 36, and the adjacent side has length 12, $\tan R = \frac{36}{12} = 3$. So angle R is approximately 71.6°. Choice (D) is the result of taking $\frac{12}{36}$ to be the cosine of R.

7. A

The measurements given allow you to apply the law of sines. First, find the measure of the angle opposite the side of length 18. You can then find the measure of the third angle.

Let $A = 32°$, $a = 12$, and $b = 18$. Since $\frac{\sin A}{a} = \frac{\sin B}{b}$, $\frac{\sin 32°}{12} = \frac{\sin B}{18}$, and $\sin B = \frac{3\sin 32°}{2} \approx \frac{3(0.5299)}{2} \approx 0.7949$. So $B \approx 52.6°$.

The degree measure of angle C, then, is $180 - (32 + 52.6) \approx 95.4$. We now have enough values to plug into $\frac{\sin A}{a} = \frac{\sin C}{c}$:

$$\frac{\sin 32°}{12} = \frac{\sin 95.4}{c}$$

$$\frac{0.5299}{12} \approx \frac{0.9956}{c}$$

$$c \approx \frac{12(0.9956)}{0.5299} \approx 22.5$$

8. A

Before using the law of sines formula, you need to find the measure of angle C. Its measure is the difference between $180°$ and the sum of the other two angle measures: $180 - (45 + 72) = 63$. So,

$$\frac{\sin 45°}{a} = \frac{\sin 63°}{24}$$

$$a = \frac{24 \sin 45°}{\sin 63°} \approx \frac{24(0.7071)}{0.8910} = 19.0$$

Choice (C), 25.6 is the length of the side opposite angle B.

9. C

Let the side with length $\sqrt{3}$ be side a, and let the side with length $\sqrt{27}$ be side b. Then $c = x$, and $C = 60°$. To find the value of x, we can plug the values we have into the law of cosines formula:

$$c^2 = a^2 + b^2 - 2ab \cos C$$
$$c^2 = \left(\sqrt{3}\right)^2 + \left(\sqrt{27}\right)^2 - 2\left(\sqrt{3}\right)\left(\sqrt{27}\right)\cos 60°$$
$$c^2 = 3 + 27 - 2\left(\sqrt{3}\right)\left(\sqrt{27}\right)0.5$$
$$c^2 = 30 - \sqrt{81} = 30 - 9 = 21$$

So, $c^2 = 21$, and $c = \sqrt{21}$.

10. E

Let $a = 43$, $b = 61$, and $c = 68$. Plugging those values into the law of cosines formula $a^2 = b^2 + c^2 - 2bc \cos A$, has this result:

$$43^2 = 61^2 + 68^2 - 2(61)(68)\cos A$$

$$1{,}849 = 3{,}721 + 4{,}624 - 8{,}296 \cos A$$

$$8{,}296 \cos A + 1{,}849 = 8{,}345$$

$$8{,}296 \cos A = 6{,}496$$

$$\cos A = \frac{6{,}496}{8{,}296} \approx 0.78303$$

$$A = 38.46°$$

CHAPTER 9

Linear Transformations

WHAT ARE LINEAR TRANSFORMATIONS?

Linear transformations are operations performed on points on the coordinate plane. They literally move points from one location to another. In Algebra II, you are likely to encounter transformations applied to individual points as well as sets of points, including lines, line segments, and geometric figures.

Linear transformations might be represented graphically or just in terms of numbers and coordinates. The techniques we'll review apply to both situations.

CONCEPTS TO HELP YOU

The three most common kinds of transformations in Algebra II are *reflections*, *rotations*, and *translations*. We'll explain the effect of each kind of transformation, and how each one is typically described.

i. Reflections

A set of points can be reflected over a line, such that the resulting set is a mirror image of the original. This line, not surprisingly, is called the *line of symmetry*. Each point in the reflection is the same distance from the line of symmetry as the corresponding point in the original set.

If the set includes multiple points, then the reflection might have a different orientation. A line or line segment, for instance, might have a different slope after it is reflected. How it is oriented depends on the line of symmetry. The line can be one of the axes on the coordinate plane, or a line parallel to an axis. It can also be a diagonal line.

The figure in the upper left quadrant of the coordinate plane is being reflected over the line $y = x$:

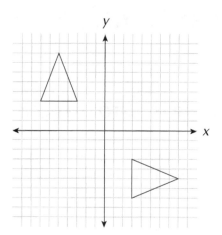

The line $y = x$ is frequently used as a line of symmetry in Algebra II questions. The x and y axes are often used as well. You will also see other horizontal and vertical lines, which are usually given in the form of equations such as $x = h$ or $y = k$, where h and k are constants.

ii. Rotations

A rotation moves an object around a given point on the coordinate plane. Most typically, that central point is the origin. However, an Algebra II question might use a different point for the object to be rotated around.

Most rotations in Algebra II involve standard degree measures, including 90° and 180°. Below, the point M is rotated 180° around the origin. The result of the rotation is the point M'.

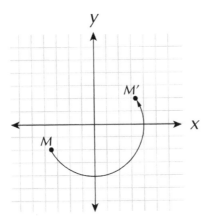

When an object containing multiple points is rotated, each point takes its own path around the central point. Therefore, like many reflections, a rotation changes the orientation of the figure or line.

iii. Translations

Translations are also known as *slides*. Translating an object is literally a matter of sliding it along the coordinate plane, as determined by a given ordered pair. If you are given a point P at coordinates (a, b), you may be asked to translate it by (c, d). Here, the ordered pair (c, d) is not a set of coordinates like (a, b). Rather, the numbers indicate the direction and distance that the point P will be moved along the plane. The coordinates of the translated point P' are $(a + c, b + d)$. That is, the x coordinate will change by c units, and the y coordinate will change by d units.

Here is the point Q, translated by $(4, -5)$. The point is moved four units to the right and five units down. The result is the point Q'.

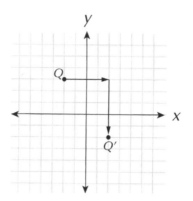

iv. Compound Transformations

Not every transformation question will involve just one reflection, rotation, or translation. You are likely to find situations where more than one transformation is carried out on a given object. It could be a series of transformations of the same kind, or it could be a combination of different ones. There is nothing new to apply in the case of compound transformations; carrying them out is just a matter of carrying out each individual transformation, one at a time.

STEPS YOU NEED TO REMEMBER

Now that we've introduced these three common kinds of transformations, we'll show you rules and formulas you can use to carry them out. You should practice with each one, following the steps we present to transform points or simple figures.

i. Finding the Coordinates of Reflected Points

To find the reflection of a set of points, start by working with one point at a time.

Let's take the point (a, b). Here are the results of reflecting over several lines of symmetry:

x–axis: $(a, -b)$

y–axis: $(-a, b)$

line $x = c$: $(2c - a, b)$

line $y = d$: $(a, 2d - b)$

line $y = x$: (b, a)

line $y = -x$: $(-b, -a)$

For other lines, you can take these steps to find the reflection P' of a point P:

- Plot the line of symmetry
- Plot a line segment perpendicular to the line of symmetry, with P as an endpoint. The segment has a length such that it intersects the line of symmetry at its midpoint. The other endpoint is P'.

Here is a point reflected over the line $y = \dfrac{2x}{3} - 1$

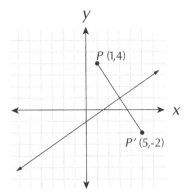

If you are reflecting a line or line segment, you might be able to complete the graph of a reflection after you've reflected a couple of key points. If you are reflecting a polygon, you can reflect the vertices, and then connect them to show the reflected sides.

ii. Finding the Coordinates of Rotations

Rotations of 90° and 180° are by far the most common in Algebra II. For a point P with coordinates (a, b), the following relations apply:

- P rotated 180° around the origin has coordinates $(-a, -b)$
- P rotated 90° clockwise around the origin has coordinates $(b, -a)$
- P rotated 90° counterclockwise around the origin has coordinates $(-b, a)$

Rotating an object around a point other than the origin is not quite as simple. You'll need to take a few extra steps. The basic idea is that you need to treat the central point as an "adjusted origin."

- Find the location of the point you are rotating relative to the central point. To do that, you subtract the coordinates of the central point from the coordinates of the point to be rotated.
- Next, rotate the "adjusted point" around the origin, according to the rules above.
- Once you have the coordinates of the rotation, add the coordinates of the central point to it. That gives you the coordinates of the object rotated around the central point, and you're done.

You'll find an illustration of this in **Question 3** below.

Remember that if the rotated object has multiple points, each one must be rotated in the same manner. If you are asked to rotate a polygon, you may need to rotate each vertex. You can then complete the rotation by connecting the vertices as they were in the original figure.

iii. Finding the Coordinates of Translations

Translations tend to be the most straightforward of the transformations. That's because carrying out a translation is just a matter of addition. As we explained earlier, if you have the point P with coordinates (a, b), then the coordinates of the point translated by (c, d) are $(a + c, b + d)$. If c is positive, then the x-coordinate of the translated point will be greater. If c is negative, that coordinate value will decrease. The same relationship holds for d and the y-coordinate.

If you are translating a geometric figure, you might find it easier to use matrix addition, which we covered in Chapter 5. It just provides a more efficient way of adjusting a number of pairs of coordinates. Suppose you have a triangle with vertices at (j, k), (l, m), and (n, o). You can represent those ordered pairs with a matrix:

$$\begin{bmatrix} j & l & n \\ k & m & o \end{bmatrix}$$

Each column has the coordinates of one vertex. Now, suppose you need to translate the triangle by (c, d). You can determine the coordinates of the translated triangle by finding this sum:

$$\begin{bmatrix} j & l & n \\ k & m & o \end{bmatrix} + \begin{bmatrix} c & c & c \\ d & d & d \end{bmatrix} = \begin{bmatrix} j+c & l+c & n+c \\ k+d & m+d & o+d \end{bmatrix}$$

Once you've evaluated each element of this last matrix, you can take each column to get the coordinates of a translated vertex.

iv. Compound Transformations

The key to carrying out compound transformations is taking it one step at a time. Identify the individual transformations specified in a question, and be sure to follow the specified order. Once you have that squared away, be sure to keep careful track of the result of each step. An incorrect transformation at any stage will send you permanently off track.

STEP-BY-STEP ILLUSTRATION OF THE FIVE MOST COMMON QUESTION TYPES

Now it's time to apply the techniques and rules for carrying out transformations. While some questions have you perform transformations, you will also be asked to compare transformations to original objects to determine what operation produced the transformation.

Question 1: Reflection Over an Axis

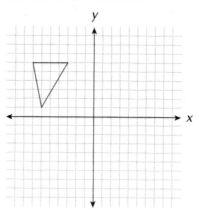

Which of the following is the reflection of the above figure across the *y*-axis?

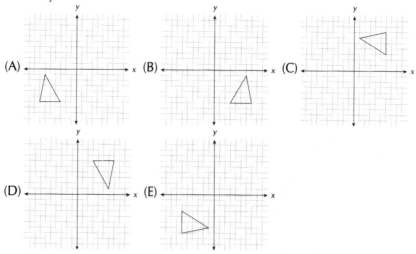

To reflect a point over the *y*-axis, you take the negative of the *x*-coordinate. To reflect a polygon such as the one we are given in this question, we can reflect each of the vertices. That will give us enough information to determine the reflection of the whole figure.

So, let's identify the vertices as *A*, *B*, and *C*, and label their coordinates.

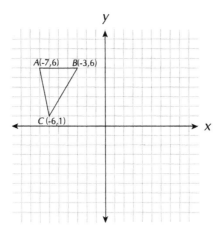

A is at (–7, 6), *B* is at (–3, 6), and *C* is at (–6, 1). The reflected vertices will be *A'*, *B'*, and *C'*. We can get those coordinates by finding the negative of the *x*-coordinate of each original point.

$A\ (–7, 6) \rightarrow A'\ (7, 6)$

$B\ (–3, 6) \rightarrow B'\ (3, 6)$

$C\ (–6, 1) \rightarrow C'\ (6, 1)$

Finally, let's plot these new points, and connect them to make the reflected triangle:

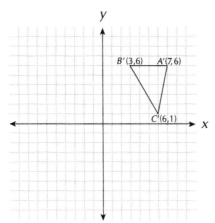

Choice (D) is the correct answer, since that graph matches the one above.

Choice (A) is the reflection of the figure over the *x*-axis. Keep in mind that a reflection over the *y*-axis affects the *x*-coordinates, but not the *y*-coordinates. Choice (C) has the transformation in the correct quadrant of the grid, but it is actually a 90° clockwise rotation of the figure, rather than a reflection.

Question 2: Reflections Over Other Lines

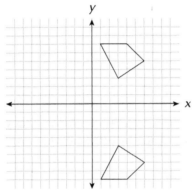

The two figures on the above grid are reflections over what line?

(A) $x = 1$

(B) $x = 2$

(C) $y = -2$

(D) $y = -1$

(E) $y = 1$

This question differs from Question 1 not just because it involves a reflection across a line other than an axis. It also shows you the reflected figure, and asks you what reflection would get it. Thus, it is a sort of reversal of Question 1. To find the line of symmetry, pick two corresponding points. Pick a vertex of the top figure, and the reflection of that point. Let's start with the top figure, and take the vertex at (3, 3); we'll label that point *P*. The corresponding point on the bottom figure, *P'*, is the point at (3, −5).

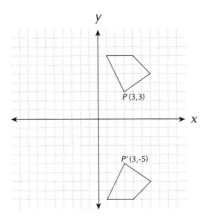

The line $y = -1$ is midway between *P* and *P'*.

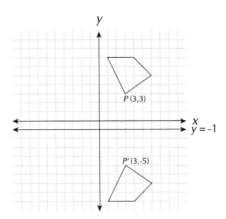

Since both points are the same distance from the line $y = -1$, they are reflected over that line, along with the other corresponding pairs of points. **Choice (D) is the correct answer.** For choice (C) to be correct, *P'* would have to be at (3, −7) instead of (3, −5). For choice (E) to be correct, *P'* would have to be at (3, −1).

Question 3: Rotations

A triangle on the coordinate plane has a vertex with coordinates (5, –2). Which of the following are coordinates of a vertex of the triangle after it is rotated 180° around the point with coordinates (–2, 4)?

(A) (–9, 10)

(B) (–7, 6)

(C) (–1, 4)

(D) (3, 2)

(E) (7, 8)

This is one of the more difficult kinds of rotation questions, since it uses a central point other than the origin. To rotate the vertex in question around this other point, we can recenter it by treating the point (–2, 4) as the origin. We subtract that pair of coordinates from (5, –2) to get the adjusted coordinates. The result is (5–(–2), –2 –(4)), or (7, –6).

These new coordinates represent the position of the point (5, –2) in relation to (–2, 4). Now we can rotate the point (7, –6) 180° around the origin. The coordinates of the rotated point are the negatives of the original coordinates, or (–7, 6).

Finally, we have to readjust the coordinates of the rotated point, since we weren't really rotating the point at (7, –6). Because we subtracted (–2, 4) in our first step, we must now add those coordinates to (–7, 6). The result is (–9, 10), and **choice (A) is the correct answer.**

Choice (B) is the result of leaving out that last step. Choice (D) is the result of adding (–2, 4) to (5, –2) in the first step instead of subtracting.

Question 4: Translations

In the coordinate grid below, *T'* is a translation of *T*.

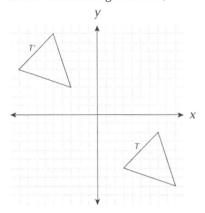

What translation of *T* results in *T'*?

(A) (–16, 7)

(B) (–12, 11)

(C) (–10, 5)

(D) (–6, 9)

(E) (–5, 13)

One thing to keep in mind as you tackle this sort of question is that each point in triangle *T* is translated in the exact same way. If we determine how a single point on *T* was translated, then we have determined how the whole figure was translated. So we can identify the translation performed here by identifying a point on *T* with its counterpart on *T'*. When dealing with a polygon, the best choice is a pair of corresponding vertices. Pairs of corresponding vertices are labeled on the following page.

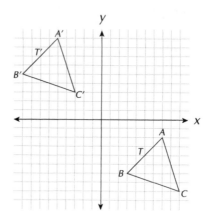

Since a translation doesn't change the orientation of a figure, we can be sure that the highest vertices are corresponding points. A has coordinates $(7, -2)$, and A' has coordinates $(-5, 9)$.

So, if A was translated by (x, y), then $7 + x = -5$, and $-2 + y = 9$. Solving those equations gets us $x = -12$ and $y = 11$, making the translation $(-12, 11)$. **Choice (B) is the correct answer.**

Choice (A) is the result of taking A and B' to be corresponding vertices. Choice (C) is the result of pairing A and C'. Choice (D) is the result of pairing B and C'.

Question 5: Compound Transformations

Which of following are the coordinates of the point at (–5, 8) rotated counterclockwise around the origin 90° and then reflected over the *x*-axis?

(A) (–8, –5)

(B) (–8, 5)

(C) (–5, –8)

(D) (5, –8)

(E) (8, –5)

This question requires you to perform two different kinds of translations. The order in which you perform them matters here, as we'll see.

To rotate any point 90° counterclockwise around the origin, you switch the coordinates, and take the negative of the new *x*-coordinate. So, the 90° rotation of (–5, 8) gets us (–8, –5).

The next step is to reflect the point over the *x*-axis. To do that, we simply take the negative of the *y*-coordinate. That gets us (–8, 5) as the result of the compound transformation, and **choice (B) is the correct answer.**

Had we performed the transformations in the opposite order, we would have gotten choice (E). Reflecting (–5, 8) over the *x*-axis would get us (–5, –8). Rotating that point 90° counterclockwise around the origin would then get us (8, –5).

CHAPTER QUIZ

1. Which of the following pairs of points are reflections of each other over the *x*-axis?

    (A) (–5, 7) and (5, –7)

    (B) (–2, –6) and (2, –6)

    (C) (3, 1) and (1, 3)

    (D) (6, 4) and (–6, –4)

    (E) (8, 9) and (8, –9)

2. Point *A* has coordinates (–5, –2). *A*′ is the reflection of *A* over the *y*-axis. What are the coordinates of *A*?

    (A) (–5, 2)

    (B) (–2, –5)

    (C) (2, 5)

    (D) (5, –2)

    (E) (5, 2)

3. What are the coordinates of the reflection of point *P* over the line shown?

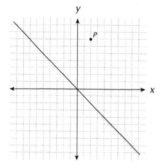

    (A) (–7, –2)

    (B) (–7, 2)

    (C) (–2, –7)

    (D) (2, –7)

    (E) (7, 2)

4. Which of the following points do you get by rotating the point at (7, –8) 180° around the origin?

   (A) (–8, 7)

   (B) (–7, –8)

   (C) (–7, 8)

   (D) (8, –7)

   (E) (8, 7)

5. Point G′ has coordinates (1, –6). It is the rotation of point G 90° counterclockwise around the origin. What are the coordinates of G?

   (A) (–6, –1)

   (B) (–6, 1)

   (C) (–1, –6)

   (D) (–1, 6)

   (E) (6, 1)

6.

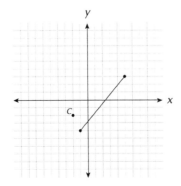

   Which of the following is the rotation of the line segment shown above 90° counterclockwise around point C?

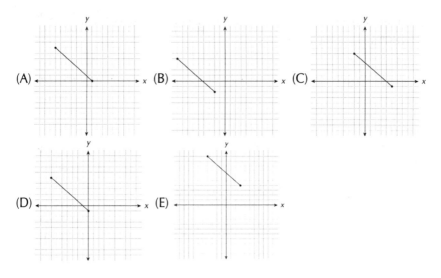

7. If translating the point at (–6, 10) on the coordinate grid by (x, y) results in a point located at (3, –20), then what are the values of x and y?

    (A) (–18, –200)

    (B) (–0.5, –2)

    (C) (–0.5, –0.5)

    (D) (9, –10)

    (E) (9, –30)

8. Which of the following transformations of point K at (5, 9) on a coordinate grid results in the point (–9, –5)?

    (A) Reflection over the line y = x

    (B) Reflection over the line y = –x

    (C) Rotation of 90° counterclockwise around the origin

    (D) Rotation of 180° around the origin

    (E) Translation of (–14, 14)

9. A line segment has endpoints at (7, −6) and (−4, −9). If the line segment is first translated by (5, −2) and then by (−8, −3), then which of the following will be the new endpoints?

(A) (4, −7) and (−11, −14)

(B) (4, −8) and (−9, −14)

(C) (4, −9) and (−9, −14)

(D) (4, −11) and (−7, −14)

(E) (4, −12) and (−6, −14)

10.

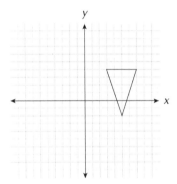

Which of the following is the transformation of the above figure, rotated counterclockwise 90° around the origin, reflected over the line x = 2, and then translated by (−3, 1)?

 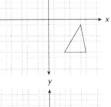 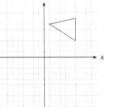

(A)   (B)   (C)

 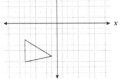

(D)   (E)

ANSWER EXPLANATIONS

1. E

Two points are reflections of each other over the x-axis if their x-coordinates are the same, and one y-coordinate is the negative of the other. Only the pair (8, 9) and (8, –9) meets that requirement. The points in (A) are reflections over the line $y = -x$, and the points in (B) have the y-axis as the line of symmetry. The line of symmetry between the points in (C) is $y = x$.

2. D

Since the reflection of a point (a, b) over the y-axis is $(-a, b)$, A' has coordinates (5, –2). Choice (A) gives the coordinates of the reflection of A over the x-axis. Choice (E) gives the coordinates of the rotation of the point 180° around the origin.

3. A

Point P has coordinates (2, 7). The line of reflection has the equation $y = -x$. Since the reflection of (a, b) over that line is $(-b, -a)$, the reflection of P has the coordinates (–7, –2). Choice (C) gives the coordinates of a 180° rotation around the origin. Choice (E) gives the coordinates of the reflection over $y = x$ instead of $y = -x$.

4. C

To find the coordinates of a 180° rotation around the origin, multiply each of the original coordinates by –1. That makes the x-coordinate –7, and the y-coordinate 8. The coordinates in choice (A) would not be the result of a rotation, but of a reflection over the line $y = x$. Choice (E) gives the coordinates of a 90° counterclockwise rotation.

5. A

Since the rotation of a point 90° counterclockwise around the origin has coordinates $(-b, a)$, say that $-b = 1$ and $a = -6$. So, $b = -1$. G has coordinates (a, b) or (–6, –1). Choice (E) would be the coordinates of G if the rotation were clockwise rather than counterclockwise.

6. D

The line segment has endpoints at (–1, –4) and (5, 3), and the central point *C* has the coordinates (–2, –2). So, the adjusted endpoints are (1, –2) and (7, 5). Rotated 90° about the origin, those points become (2, 1) and (–5, 7). Readjusting those coordinates by adding those of the central point gets you (0, –1) and (–7, 5). Those coordinates match the points in choice (D).

7. E

Since the point (–6, 10) is translated by (*x*, *y*) to get (3, –20), –6 + *x* = 3 and 10 + *y* = –20. There, *x* = 9, *y* = –30, and the translation is (9, –30). Choice (A) is the result of multiplying each pair of corresponding coordinates. Choice (B) gives the solutions to –6*x* = 3, and 10*y* = –20 instead of –6 + *x* = 3, and 10 + *y* = –20.

8. B

If the coordinates of *K* are (*a*, *b*), then the coordinates of the transformation are (–*b*, –*a*), since the *x*- and *y*-coordinates are reversed and multiplied by –1. That matches the result of a reflection over the line *y* = –*x*.

The reflection of *K* over *y* = *x* has the coordinates (9, 5).

The 90° counterclockwise rotation of *K* results in a point at (–9, 5).

The 180° rotation of *K* results in a point at (–5, –9).

The translation of *K* by (–14, 14) has the coordinates (–9, 23). A translation of (–14, –14) would get the desired coordinates.

9. D

You can use matrices to handle compound translation questions more easily.

The matrix $\begin{bmatrix} 7 & -4 \\ -6 & -9 \end{bmatrix}$ represents the endpoints of the line segment. The

matrices $\begin{bmatrix} 5 & 5 \\ -2 & -2 \end{bmatrix}$ and $\begin{bmatrix} -8 & -8 \\ -3 & -3 \end{bmatrix}$ represent the two translations.

Since $\begin{bmatrix} 5 & 5 \\ -2 & -2 \end{bmatrix} + \begin{bmatrix} -8 & -8 \\ -3 & -3 \end{bmatrix} = \begin{bmatrix} -3 & -3 \\ -5 & -5 \end{bmatrix}$ you can use that sum as the lone

translation matrix: $\begin{bmatrix} 7 & -4 \\ -6 & -9 \end{bmatrix} + \begin{bmatrix} -3 & -3 \\ -5 & -5 \end{bmatrix} = \begin{bmatrix} 4 & -7 \\ -11 & -14 \end{bmatrix}$

That sum represents the points $(4, -11)$ and $(-7, -14)$.

Be careful to read the coordinates of the matrix carefully. Choice (A) is the result of using the rows instead of the columns to get the pairs of the coordinates. Choice (C) results from putting the original coordinates in rows instead of columns, such that you might use

$\begin{bmatrix} 7 & -6 \\ -4 & -9 \end{bmatrix} + \begin{bmatrix} -3 & -3 \\ -5 & -5 \end{bmatrix} = \begin{bmatrix} 4 & -9 \\ -9 & -14 \end{bmatrix}$ to find the new endpoints.

10. C

This compound transformation involves three transformations.

The first transformation, a rotation, results in this figure:

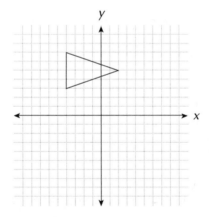

Reflecting that figure over the line $x = 2$ gives you this:

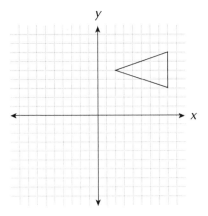

Finally, translating the above figure by $(-3, 1)$ gives you this:

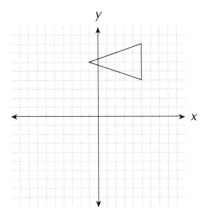

The final result matches the graph in choice (C).

The figure in choice (A) is the result of starting with a clockwise rotation instead of a counterclockwise one. You might notice that choice (B) is the figure we got in the second step above; the last step was simply not performed on that figure. Choice (D) shows the figure you would get by using $y = 2$ instead of $x = 2$ as the line of reflection. Choice (E) is the result of leaving out the first step, the counterclockwise rotation.

Sequences and Series

WHAT ARE SEQUENCES AND SERIES?

A *sequence* is an ordered list of numbers. Each number in a sequence is an *element* or *term*. Although the numbers of a sequence have an order, the list need not follow a rule or pattern. However, in Algebra II, our focus is on two kinds of sequences that follow certain algebraic patterns. They are *arithmetic sequences* and *geometric sequences*.

Some sequences have a set number of terms. Others go on forever, following a basic pattern.

A *series* is the sum of the members of a sequence. While words *sequence* and *series* might be used as synonyms elsewhere, they have very precise meanings in algebra. Be careful not to confuse them.

CONCEPTS TO HELP YOU

Here, we'll further explore the concepts of sequences and series. You'll see that one of their most important features is the relation between consecutive terms.

i. Sequence Notation

A sequence can be identified with a single term, such as A or A_n. You can identify a term in a sequence as a_i, where i is the place of the term in the series. For instance, the second term in a sequence is a_2, and the tenth term is a_{10}.

ii. Arithmetic Sequences

The terms of an arithmetic sequence increase or decrease by a fixed amount. That fixed amount is called the *common difference*, which is usually represented with the variable d. Each term in an arithmetic sequence is the sum of the previous term and the common difference.

So, in an arithmetic sequence, $a_2 = a_1 + d$, $a_3 = a_2 + d$, and so on. Generally, $a_i + d = a_{i+1}$. Notice also that since $a_2 = a_1 + d$ and $a_3 = a_2 + d$, $a_3 = a_1 + 2d$.

The sequence 2, 6, 10, 14, 18 is an arithmetic sequence. In that sequence, $d = 4$. If you subtract any term from the one that follows it, you'll get a difference of 4.

On the other hand, the sequence 4, 8, 11, 15 is not an arithmetic sequence. The difference between a_1 and a_2 is 4, but the difference between a_2 and a_3 is 3.

The sequence 2, 6, 10, 14, 18, ... has a few things in common with 2, 6, 10, 14, 18. Each has the same value for a_1 and d. This sequence is different because it is infinite. That it ends with the comma and dots indicates that it goes on and on. Unlike the sequence 2, 6, 10, 14, 18, the sequence 2, 6, 10, 14, 18, ... also has the elements 22, 26, 30, and so on.

iii. Geometric Sequences

In a geometric sequence, the ratio of any term to the next one is constant. The ratio of the first term to the second term is the same as that of the second to the third, and so on. This number, often represented by the variable r, is known at the *common ratio*.

In a geometric sequence, $a_2 = a_1 r$, $a_3 = a_2 r$, and so on. In general, $a_i r = a_{i+1}$. Since $a_2 = a_1 r$ and $a_3 = a_2 r$, $a_3 = a_1 r^2$.

iv. Series

A series is the sum of the terms of a sequence. As long as the sequence is finite, the series can be calculated. An arithmetic series of an infinite sequence cannot be calculated. You can find the sum of a limited part of an infinite arithmetic series, however, and many Algebra II questions will ask you to do just that.

The same goes for geometric series, with one exception. If the common ratio is less than 1 and greater than –1, then the terms in a sequence get closer and closer to 0. As a result, the sum of the terms in the sequence approaches a limit, and has a finite sum.

The term S_n is often used to represent the first n terms of a series. The Greek letter Σ (sigma) is also used to indicate series. The sum of the ith through jth terms of a series is represented by the expression $\sum\limits_{n=i}^{j} a_n$.

This expression has the value of $a_i + a_{i+1} + ... + a_j - 1 + a_j$.

Let's take the series 2, 7, 12, 17, 22, ... Here,

$$S_4 = \sum\limits_{n=1}^{4} a_n = 2 + 7 + 12 + 17 = 38 .$$

STEPS YOU NEED TO REMEMBER

This section will give you the tools you need to tackle virtually any sequence or series question. It all boils down to a few basic formulas. You can use them to find unknown terms in a sequence, common differences or ratios, and sums of series.

i. Identifying a Sequence

Some questions will present a sequence without specifying which kind it is, arithmetic or geometric. Before you go about setting up the solution to such a question, you'll want to be certain about which kind you're dealing with. Now, if you're not told whether the sequence in question is arithmetic or geometric, then that can't very well specify a common difference or common ratio. Instead, if the question is to give you enough to work with, it will have to give you the values of at least three terms. You'll have to determine whether those terms, given their places in the sequences, fit a pattern of terms that has a common difference or a common ratio.

ii. Working with Arithmetic Sequences

The following is the most important formula for setting up solutions to arithmetic sequence questions.

If a_i and a_j are terms in an arithmetic sequence, and $j > i$, then

$a_j = a_i + (j - i)d.$

So, for instance, $a_j = a_1 + (j - 1)d.$

If you know the values of two terms in an arithmetic sequence (and their places in the sequence, you can plug them into the formula and solve it to find the common difference. Likewise, if you know the common difference and the value of one term of such a sequence, you can use the formula to find the value of another term.

iii. Working with Geometric Sequences

In a geometric sequence where r is the ratio between consecutive terms, a_1 is the first term, and a_n is the nth term, then $a_n = a_1 r^{n-1}$.

For instance, if the first term in a series is 3 and the ratio is 5, then the fourth term is

$a_5 = 3(5)^{4-1} = 3(5)^3 = 3(125) = 375$

More generally, if $j > i$, then $a_j = a_i r^{j-i}$.

So, if you know the value of two of the three unknowns in this formula, a_i, a_j, and r, then you can plug them into the formula, and solve it for the third one.

Solving such an equation might require you to find the root of a number. Since $a_n = a_1 r^{n-1}$, $r^{n-1} = \dfrac{a_n}{a_1}$. Once you have the value of $\dfrac{a_n}{a_1}$, you'll have to find its $(n-1)^{th}$ root to get the value of r.

iv. Finding a Part of a Series

The sum of the first n terms of an arithmetic series is given by this formula:

$$S_n = \frac{n}{2}(a_1 + a_n).$$

So, the sum of the first three terms of $3, 6, 9, 12, 15, \ldots$ is

$S_3 = \frac{3}{2}(3 + 9) = \frac{3}{2}(12) = 18$. You can verify that by finding that $3 + 6 + 9 = 18$.

The sum of the first n terms of a geometric series is given by this formula:

$$S_n = \frac{a_1 - a_1 r^n}{1 - r}.$$

The sum of the first three terms of $3, 6, 12, 24, 48, \ldots$ is

$$S_3 = \frac{3 - 3(2^3)}{1 - 2} = \frac{3 - 3(8)}{-1} = \frac{3 - 24}{-1} = \frac{-21}{-1} = 21.$$

This shows that $3 + 6 + 12 = 21$.

Suppose that, rather than being asked to find the sum of the first n terms of a series, you are asked to find the sum of the ith through jth terms of a series. That sum is the value of $S_j - S_{i-1}$. For example, the sum of the third through seventh terms of a series is the sum of the first seven terms, minus the sum of the first two terms.

v. Finding the Sum of an Infinite Geometric Series

As we explained earlier, an infinite geometric series has a sum if $-1 < r < 1$. If you know the value of r and the first term, then you can use this formula to find the sum:

$$S = \frac{a_1}{1 - r}$$

For example, in the series $2 + 1 + 0.5 + 0.25 + 0.125 + \ldots, a_1 = 2$ and $r = 0.5$. So, $S = \frac{a_1}{1 - r} = \frac{2}{1 - 0.5} = \frac{2}{0.5} = 4$.

If that doesn't seem possible to you at first glance, consider these sums:

$S_2 = 2 + 1 = 3$	$S_6 = 3.875 + 0.0625 = 3.9375$
$S_3 = 3 + 0.5 = 3.5$	$S_7 = 3.9375 + 0.03125 = 3.96875$
$S_4 = 3.5 + 0.25 = 3.75$	$S_8 = 3.96875 + 0.015625 = 3.984375$
$S_5 = 3.75 + 0.125 = 3.875$	$S_9 = 3.984375 + 0.0078125 = 3.9921875$

You could continue to add more terms, and the sum would get ever closer to 4, without coming to exactly 4. Since the series is infinite, we say that the sum is exactly 4.

STEP-BY-STEP ILLUSTRATION OF THE FIVE MOST COMMON QUESTION TYPES

Now let's apply what we've just reviewed to set up the solutions to some questions. You'll find that the handful of formulas we just introduced can be adapted for a number of purposes.

Question 1: Finding the Next Term in a Sequence

What is the next term in the sequence 4, 20, 100, ...?

(A) 124

(B) 180

(C) 500

(D) 2,000

(E) 2,500

Our first step must be to find out what kind of sequence we are dealing with. Is it arithmetic or geometric? Since $20 - 4 = 16$, and $100 - 20 = 80$, there is no common difference; however, since $\frac{20}{4} = \frac{100}{20} = 5$, the terms increase by a common ratio. That makes the sequence a geometric one. The common ratio is 5, so each term in the sequence is the product of the previous term and 5. The product of 100 and 5 is 500. You can also use the geometric sequence formula to find the fourth term:

$$a_n = a_4 = a_1 r^{n-1} = 4(5^{4-1}) = 4(5^3) = 4(125) = 500.$$

Choice (C) is the correct answer.

Choice (B) is the result of treating this as an arithmetic sequence, and adding 80, the difference between 100 and 20, to 100. Choice (E) is 100 multiplied by 25. You might get that value if you determined the ratio by dividing a_3 by a_1 instead of a_2.

Question 2: Finding a Missing Term in a Sequence

If the second term in an arithmetic sequence is 10 and the eighth term is 34, then what is the sixth term?

(A) 22
(B) 26
(C) 28
(D) 30
(E) 32

This question provides the values of a_2 and a_8, and asks for the value of a_6. To get that value, you will first need to find the common difference d in this arithmetic sequence. To find it, solve $a_j = a_i + (j - i)d$ for d, where $a_i = 10$, $a_j = 34$, $i = 2$, and $j = 8$:

$$34 = 10 + (8 - 2)d$$
$$24 = (8 - 2)d$$
$$24 = 6d$$
$$d = 4$$

Now we can use the formula again, plugging in the values of d and a_2 to find a_6.

$$a_6 = a_2 + (6 - 2)d = 10 + (6 - 2)4 = 10 + (4)4 = 10 + 16 = 26.$$

So, choice (B) is the correct answer. Choice (A) is the value of a_5, and (D) is the value of a_7.

Question 3: Finding the Sum of a Part of an Arithmetic Series

If $A_n = 13, 7, 1, -5, \ldots$, then what is the sum of the first seven numbers of the series?

(A) −70

(B) −56

(C) −45

(D) −35

(E) −23

This question asks you to find the value of S_7. For that, you will need to find the value of the seventh term, a_7, since that is one of the terms in the formula $S_n = \dfrac{n}{2}(a_1 + a_n)$.

To find that term, use the formula $a_n = a_1 + (n-1)d$. Here $a_1 = 13$, and $d = 7 - 13 = -6$. So, $a_7 = 13 + (7-1)(-6) = 13 + 6(-6) = 13 - 36 = -23$.

Since $a_7 = -23$, $S_7 = \dfrac{7}{2}(13 + (-23)) = \dfrac{7}{2}(-10) = -\dfrac{70}{2} = -35$, and **choice (D) is the correct answer.**

Choice (E) is just the seventh term in the sequence. You need that to get the solution, but it is not the last step.

Question 4: Finding the Sum of a Part of a Geometric Series

For the geometric sequence −2, 6, −18, ... what is the value of

$$\sum_{n=5}^{8} a_n ?$$

(A) −9,720

(B) −4,374

(C) −1,080

(D) 2,916

(E) 3,240

This question calls for the sum of the fifth, sixth, seventh, and eighth terms in the sequence. To find it, you can find the difference between the sum of the first eight terms and the sum of the first four terms. Since $a1 = -2$ and $r = -3$,

$$S_8 = \frac{-2 - (-2)(-3)^8}{1 - (-3)} = \frac{-2 - (-2)(6,561)}{4} = \frac{-2 + 13,122}{4} = \frac{13,120}{4} = 3,280,$$

and $S_4 = \dfrac{-2 - (-2)(-3)^4}{1 - (-3)} = \dfrac{-2 - (-2)(81)}{4} = \dfrac{-2 + 162}{4} = \dfrac{160}{4} = 40.$

$S_8 - S_4 = 3,280 - 40 = 3,240.$

So, choice (E) is the correct answer.

Choice (A) is the sum of the sixth through ninth terms. Choice (C) is the sum of the fourth through seventh terms.

Question 5: Finding an Infinite Geometric Series

The second term of a geometric sequence is 12 and the third term is 4. What is the value of the series?

(A) 18

(B) 36

(C) 48

(D) 54

(E) 72

This question gives us the values of two consecutive terms in a geometric series. We'll have to use them to find the first term, which is needed for the series formula. Since the common ratio is the value of any term divided by the previous one, $r = \dfrac{4}{12} = \dfrac{1}{3}$ (since $-1 < \dfrac{1}{3} < 1$, we can be sure that this infinite geometric series has a sum).

So, $12 = a_1\left(\dfrac{1}{3}\right)$, and $a_1 = 36$. Now we can plug our values in to find the sum:

$$S = \frac{a_1}{1 - r} = \frac{36}{1 - \dfrac{1}{3}} = \frac{36}{\left(\dfrac{2}{3}\right)} = 54. \textbf{ (D) is the correct answer choice.}$$

Choice (A) would be the sum if 12 were the first term, rather than the second. Choice (B) is just the first term.

CHAPTER QUIZ

1. Which of the following sequences is neither arithmetic nor geometric?

   (A) −7, 14, −28, 56

   (B) $1\frac{1}{2}, 1, \frac{2}{3}, \frac{4}{9}$

   (C) 5, 1.5, −2, −5.5

   (D) 6, 18, 36, 54

   (E) $8\frac{1}{3}, 10\frac{2}{3}, 13, 15\frac{1}{3}$

2. What is the next term in the sequence −6.5, −2.2, 2.1, …?

   _____

3. What is the next term in the geometric sequence −7, 14, …?

   (A) −28

   (B) −21

   (C) 21

   (D) 28

   (E) 35

4. The first term in a geometric sequence is 4 and the third term is 144. What is the fifth term?

   (A) 424

   (B) 864

   (C) 2,304

   (D) 5,184

   (E) 9,216

5. If the fifth term of an arithmetic sequence is 32 and the sixth term is 41, then what is the first term?

   (A) −4

   (B) 4

   (C) 5

   (D) 23

   (E) 77

6. What is the sum of the first six terms of the series 2.5, 7.5, 22.5, …?

   (A) 135

   (B) 607.5

   (C) 910

   (D) 1,822.5

   (E) 1,830

7. The sum of the first four terms of a geometric series is 936. The common ratio of the series is 5. What is the first term of the series?

   _____

8. The first term of an arithmetic sequence is 5, and the second term is 18. What is the sum of the 10th through the 20th terms?

   (A) 1,935
   (B) 2,057
   (C) 2,570
   (D) 3,610
   (E) 3,740

9. What is the sum of the series 200, 40, 8, ...?

   _____

10. The first term of an infinite geometric series is 24. If the sum of the series is 96, then what is the common ratio?

    (A) 0.25
    (B) 0.5
    (C) 0.75
    (D) 0.8
    (E) 1.25

ANSWER EXPLANATIONS

1. D

Choice (A) is a geometric sequence with a common ratio of -2. Choice (B) is also geometric; the common ratio is $\frac{2}{3}$. Choices (C) and (E) are both arithmetic sequences. The common difference is 3.5 in (C) and $2\frac{1}{3}$ in (E).

In choice (D), $a_2 - a_1 = 12$, and $\frac{a_2}{a_1} = 3$. However, $a_3 - a_2 = 18$, and $\frac{a_3}{a_2} = 2$. Since this series has neither a common difference nor a common ratio, it is neither arithmetic nor geometric.

2. 6.4

Since $-6.5 - (-2.2) = -2.2 - (2.1) = 4.3$, the common difference d is 4.3. The next term in the sequence, then, is $2.1 + 4.3$, which is 6.4.

3. A

Each pair of consecutive terms in a geometric sequence has a common ratio. Here, the ratio r is $\frac{14}{-7} = -2$. So, the next term is 14×-2, or -28. Choice (B) would be the third term if the common ratio were 2, rather than -2. Choice (E) would be the third term if this were an arithmetic sequence, since the difference between the first and second terms is 21.

4. D

Since $a_1 = 4$, $a_3 = 144$, and $a_3 = a_1 r^2$, $144 = 4r^2$. So, $r^2 = 36$, and r, the common ratio, is 6. Thus, $a_5 = a_3 r^2 = 144(36) = 5,184$. Choice (A) would be the fifth term if this were an arithmetic sequence. Choice (B) is the value of the fourth term in the sequence. Choice (C) uses 4 instead of 6 as the value of r, which (E) uses 8 instead of 6.

5.   A

Since $a_5 = 32$, and $a_6 = 41$, $d = 9$. You can plug the value of a_5 and d into $a_j = a_i + (j - i)d$ to find the value of a_1:

$$32 = a_1 + (5 - 1)9$$
$$32 = a_1 + 36$$
$$a_1 = -4$$

You would have gotten choice (B) if you subtracted 32 from 36 instead of 36 from 32 in the last step. Choice (C) is the second term, and choice (D) is the fourth term. Choice (E) would be first term if 41 were fifth and 32 were first, rather than the reverse.

6.   C

This is a geometric series where $r = 3 \left(\dfrac{7.5}{2.5} = \dfrac{22.5}{7.5} = 3 \right)$. Since $a_1 = 7.5$,

$$S_6 = \frac{a_1 - a_1 r^n}{1 - r} = \frac{2.5 - 2.5(3)^6}{1 - 3} = \frac{2.5 - 2.5(729)}{-2} = \frac{2.5 - 1{,}822.5}{-2} = \frac{-1{,}820}{-2} = 910.$$

Choice (A) is the product of 22.5 and 6, while (B) is just the value of the sixth term. Choice (D) is the value of $a_1 r^6$. Choice (E) is the result of using the formula for the sum of the first six elements of an arithmetic series.

7.   6

The sum on the first n terms of a geometric series is

$$S_n = \frac{a_1 - a_1 r^n}{1 - r}.$$ Here, $n = 4$, $S_n = 936$, and $r = 5$. So,

$$S_4 = 936 = \frac{a_1 - a_1 (5)^4}{1 - 5} = \frac{a_1 - 625 a_1}{-4} = \frac{624 a_1}{4} = 156 a_1.$$

So, $156 a_1 = 936$, and $a_1 = 6$.

8.   B

The sum of the 10th through the 20th terms is the sum of the first 20 terms minus the sum of the first 9 terms. Since $a_1 = 5$ and $d = 13$, $a_9 = 5 + 8(13) = 5 + 104 = 109$, and $a_{20} = 5 + 19(13) = 5 + 247 = 252$.

So, $S_9 = \frac{9}{2}(5 + 109) = \frac{9}{2}(114) = 513$, and

$S_{20} = \frac{20}{2}(5 + 252) = 10(257) = 2,570$.

$S_{20} - S_9 = 2,570 - 513 = 2,057$.

Choice (A) is the difference between S_{20} and S_{10}. Choice (C) is just the value of S_{20}. Choice (D) is the value of $\frac{n}{2}(a_9 + a_{20})$, and (E) is the value of $\frac{n}{2}(a_{10} + a_{20})$.

9. **250**

This is a geometric series where $r = \frac{1}{5}$, since $\frac{40}{200} = \frac{8}{40} = \frac{1}{5}$. $a_1 = 200$, so

$S = \frac{a_1}{1-r} = \frac{200}{1 - \frac{1}{5}} = \frac{200}{\left(\frac{4}{5}\right)} = 250$.

10. **C**

Here, you are told that $a_1 = 24$ and $S = 96$. You can plug those values into $S = \frac{a_1}{1-r}$ and solve for r:

$96 = \frac{24}{1-r}$

$1 - r = \frac{24}{96}$

$1 - r = \frac{1}{4}$

$r = \frac{3}{4} = 0.75$

Choice (A) is the result of using the formula $S = \frac{a_1}{r}$ instead of $S = \frac{a_1}{1-r}$.
Choice (E) is the result of using $S = \frac{a_1}{r-1}$ instead of $S = \frac{a_1}{1-r}$.

Combinations and Permutations

WHAT ARE COMBINATIONS AND PERMUTATIONS?

Combination and permutation questions involve counting. In particular, they require you to count a number of possible groups. They can be groups of letters, numbers, people, or any objects that can be organized, sorted, rearranged, and so on. *Combination* questions ask you to take a group of objects, and count the number of possible subgroups of a certain size. *Permutation* questions ask you to find the number of *ordered* groups of a certain size.

CONCEPTS TO HELP YOU

In many cases, it is possible to count combinations or permutations of objects by listing them. In many others, however, hundreds or even thousands of possibilities may be involved. Fortunately, you can use multiplication to count more efficiently. This multiplication often involves factorials, which we'll explain below.

i. Factorials

In how many ways can we arrange the first four letters of the alphabet? How many possible orderings of the letters *A*, *B*, *C*, and *D* are there? Given those four letters, we have four choices for the first letter of the sequence. That leaves three letters for the next place in the sequence. For the first two letters in the sequence, there are 12 possibilities:

AB	BA	CA	DA
AC	BC	CB	DB
AD	BD	CD	DC

For every first-letter selection, there are three second-letter selections. Thus, the total number of two-letter sequences is the product of 4 and 3.

Here is the complete list of possibilities for a three-letter sequence.

ABC	BAC	CAB	DAB
ABD	BAD	CAD	DAC
ACB	BCA	CBA	DBA
ACD	BCD	CBD	DBC
ADB	BDA	CDA	DCA
ADC	BDC	CDB	DCB

For every selection of the first two letters, there are two third-letter selections, since two letters are left over at that point. Thus, the total number of three-letter sequences is the product of 12 and 2, or 4, 3, and 2. That product is 24.

Once the first three letters are selected, there is only one letter remaining for the fourth spot in the order. Thus, the number of four-letter sequences is the same.

ABCD	BACD	CABD	DABC
ABDC	BADC	CADB	DACB
ACBD	BCAD	CBAD	DBAC
ACDB	BCDA	CBDA	DBCA
ADBC	BDAC	CDAB	DCAB
ADCB	BDCA	CDBA	DCBA

The number of four-letter sequences is $4 \times 3 \times 2 \times 1$. We can write that expression as 4!. That is a *factorial* expression. The factorial of a number is the product of 1 and every other integer, up to and including that number. For instance, the factorial of 6, or 6!, has a value equal to $6 \times 5 \times 4 \times 3 \times 2 \times 1$. Likewise, $8! = 8 \times 7 \times 6 \times 5 \times 4 \times 3 \times 2 \times 1$. Because 6! = 720, you would find that there are 720 ways of ordering the first six letters of the alphabet.

The factorial is the key to answering many combination and permutation questions. Rather than writing out a list of permutations or combinations and then counting up the members, as we did above, you can rely on certain formulas that use factorials to come up with a total.

0! HAS A VALUE OF 1

0! = 1, because 0! is thought to be the product of no numbers. Mathematicians treat the product of no numbers as having the value of 1, which is the product of any number and its reciprocal or multiplicative inverse.

Suppose you are asked to find the number of two-letter sequences including any of the first five letters of the alphabet (such as *BC* or *EA*). Since there are 5 possibilities for the first place and 4 for the second, the total number is 20.

How can you express that total in terms of factorials? $5! = 5 \times 4 \times 3 \times 2 \times 1$. Since $5 \times 4 = 20$, $5! = 20 \times (3 \times 2 \times 1)$. Since $3 \times 2 \times 1 = 3!$, $5! = 20 \times 3!$, and $\dfrac{20 \times 3!}{3!} = 20$. Therefore, the number of permutations of two of the first five letters of the alphabet is $\dfrac{5!}{3!}$, or 20.

Factorial expressions like $\dfrac{5!}{3!}$ can be simplified on account of the fact that numbers in the numerator and the denominator cancel each other out:

$$\frac{5!}{3!} = \frac{5 \times 4 \times 3 \times 2 \times 1}{3 \times 2 \times 1} = \frac{5 \times 4 \times \cancel{3} \times \cancel{2} \times \cancel{1}}{\cancel{3} \times \cancel{2} \times \cancel{1}} = \frac{5 \times 4}{1} = 20$$

Suppose that you are instead asked to find the number of *unordered* pairings of those first five letters of the alphabet. There are 20 ordered pairings, but for every two of those, there is an unordered pair. The set or ordered pairings contains *AB* and *BA*; the set of unordered pairings just includes the pairing *AB* (or *BA*, which is identical). So, the number of unordered pairings is half the number of ordered ones. 10 is the number of combinations of two objects from a five-object group, while 20 is the number of permutations.

In the **Steps You Need to Remember** section, we'll present useful combination and permutation formulas you can apply more generally.

ii. Other Combinations

Not every question calls for a factorial formula. Remember that factorials give you a way of counting groups when the number of remaining openings in a group depends on what has already been chosen. In the example with the first few letters of the alphabet, there are four possibilities for the second letter, since the fifth was already used in the fifth spot.

What if each letter could be used more than once? How many permutations of two letters from the first five of the alphabet are there in that case? Since there are five possibilities for the first spot and five for the second, the total is 5^2, or 25.

The general rule is that you use factorials to count combinations or permutations when the objects in question come from the same group. When the objects come from different groups, then earlier selections don't narrow down the remaining possibilities.

STEPS YOU NEED TO REMEMBER

Now we'll show you how to use multiplication systematically as a way to count combinations and permutations. Keep in mind that you need to size up a question correctly before deciding exactly how to move ahead.

i. Sizing Up the Question

Many questions in this area of Algebra II won't use the terms *combination* or *permutation*. Instead, they'll present a scenario, and it will be up to you to determine whether it calls for counting combinations or permutations.

The key to that decision often lies in whether the groups you are asked to count are supposed to be ordered. For instance, if a question asks for the number of possible selections of five books from a set of ten, then it is a combination question. On the other hand, if it asks for the total number of possible arrangements of five books, taken from a set of ten on a bookshelf, then it is asking for the number of permutations.

Alternatively, a question might actually involve permutations, rather than just combinations, if each member of the group has a particular role or function. For instance, if a question simply asks for the number of possible three-person committees that might be formed from a group of ten people, then it is a combination question. However, if it asks instead for the number of possible committees, where one person is the leader and another is the secretary, then it is a permutation question. That's because there are a number of ways to organize the same group of three people. When each place in a group is different from the others, you are dealing with permutations.

ii. Using Formulas for Permutations

Since the basic permutation formula is simpler than the basic combination formula, we'll discuss it first. As we saw in the **Concepts to Help You** section, the number of permutations of two letters from a group of five letters is 20. That's the value of 5×4, the number of possibilities for the first object in the permutations, multiplied by the number of possibilities for the second object.

In general, the number of permutations of r objects, taken from a larger group of n objects is $n \times (n-1) \times \ldots \times (n-r+1)$. Therefore, to find the number of permutations of a subgroup of r objects taken from a group of n objects, evaluate $\dfrac{n!}{(n-r)!}$. This expression is often symbolized as $P(n,r)$ or $_nP_r$.

In some instances, a group will have several identical members. For example, the digits of the number 1,231 make up a group with two identical members, since 1 appears twice. Therefore, the number of unique permutations of the digits of 1,231 is less than the number of unique permutations of four-digit numbers such as 1,234 (where each digit is unique).

Suppose you have a group of n objects, where a certain number t of the objects are identical. The number of permutations of that group is $\dfrac{n!}{t!}$. So, when you are asked to find the number of permutations of a group, you need to determine whether the objects in the group are all unique before you set up your formula.

iii. Using Formulas for Combinations

If the number of permutations of r objects, taken from a group of n objects is $P(n,r)$, then the number of combinations $C(n,r)$ will be less. For every combination of r objects, there are $r!$ ways to order it. Therefore, the number of possible r-member unordered subgroups of an n-member group is ${}_nC_r$, or $C(n,r) = \dfrac{n!}{r!(n-r)!}$.

When you are combining objects from different groups, you probably won't have to use factorials. Suppose you need to form a two-object combination, with one object from a group A, and another from a group B. The number of possible combinations is the number of objects in A times the number of objects in B.

STEP-BY-STEP ILLUSTRATION OF THE FIVE MOST COMMON QUESTION TYPES

Now let's apply these formulas and techniques to a variety of combination and permutation questions. You will find that many questions of these types are word problems. Part of the challenge of these questions lies in representing real-world problems in algebraic terms.

Question 1: Finding Combinations from a Single Group

Four students from a group of ten will be selected to serve on a special committee. How many possible different groups could be formed?

(A) 24

(B) 210

(C) 720

(D) 5,040

(E) 151,200

This question asks us to find the number of possible four-object combinations of ten different objects. We can use the combination formula to find that number. Since the larger group has ten objects and the smaller group has four, we have to find the value of $C(10,4)$. Since

$$C(n,r) = \frac{n!}{r!(n-r)!}, C(10,4) = \frac{10!}{4!(10-4)!} = \frac{10!}{4!6!} = \frac{10 \times 9 \times 8 \times 7}{4 \times 3 \times 2 \times 1} = \frac{5{,}040}{24} = 210.$$

So, there are 210 possible groups, and **choice (B) is the correct answer.**

Choice (A) is the value of 4!, and (B) is the value of $(10-4)!$, or 6!. Choice (D) is the value of $\frac{10!}{6!}$, or $\frac{10!}{(10-4)!}$. That is the number of ordered four-member groups, or permutations, of objects taken from the larger group of ten. Choice (E) is the value of $\frac{10!}{4!}$, which is actually the number of permutations of six objects, taken from a group of 10 $({}_{10}P_6)$.

Question 2: Finding Combinations from Different Groups

A cafeteria serves four different soups, three different salads, six different main courses, and four different deserts. An "Early Bird Special" meal includes one item from each group. How many different "Early Bird Special" meals are possible?

(A) 12

(B) 24

(C) 216

(D) 288

(E) 432

This question involves the combination of objects from different groups. Thus, no factorials are involved. The selection of the first object does not affect the number of objects remaining. No matter which of the four soups is selected, there are three salads to select from. So, there are 12 possible combinations of soups and salads (4×3).

The total number of possible meals, then, is the product of the number of members in each group. $4 \times 3 \times 6 \times 4 = 288$, so **(D) is the correct answer choice.**

Choice (A) might be a result of some confusion with the combination formula, because 12 is 288 divided by 4!. Choice (C) is the value of $4 \times 3 \times 6 \times 3$, and (E) is the value of $4 \times 3 \times 6 \times 6$.

Question 3: Permutations of a Whole Set

In how many different ways can the letters of the word *COMBINE* be arranged?

(A) 49

(B) 128

(C) 720

(D) 2,520

(E) 5,040

The word in question has seven letters. Therefore, we need to find the value of $_7P_7$. That represents the number of permutations of seven unique objects.

The standard permutation formula is $P(n,r) = \dfrac{n!}{(n-r)!}$. Since $n = r$ in this case, $n - r = 0$.

Recall that $0! = 1$. This means that $P(7,7) = \dfrac{7!}{0!} = 7!$.

$7 \times 6 \times 5 \times 4 \times 3 \times 2 \times 1 = 5,040$. So, there are 5,040 ways to arrange the letters of *COMBINE*, and **choice (E) is the correct answer.**

Choice (A) is 7^2, and (B) is 2^7. Choice (C) is the value of $\dfrac{7!}{7}$, the equivalent of 6!.

Question 4: Permutations of Part of a Set

A set of three different songs will be played at random from a CD with seven songs. How many possible ordered sets could be played?

(A) 35

(B) 210

(C) 343

(D) 840

(E) 1,260

Since this question calls for the number of *ordered* sets of songs, it is asking us to find the number of permutations, rather than the number of combinations. We need to find the value of $P(7,3)$, the number of permutations of three objects from a group of seven.

Since $P(n,r) = \dfrac{n!}{(n-r)!}$, $P(7,3) = \dfrac{7!}{(7-3)!} = \dfrac{7!}{4!} = 7 \times 6 \times 5 = 210$. That

is the number of possible three-song play lists, given seven total songs. Any of the seven songs can be played first. Once a song is selected for that spot, there are six remaining for the second spot, and then five for the third and last spot. **Choice (B) is the correct answer.**

Choice (A), 35, is the number of combinations of three songs from a group of seven. There are 3! or six ways to order every three-song combination, so the number of permutations, 210, is six times the number of combinations. Choice (C) is the value of 7^3. That would be the number of playlists in which any song could be repeated. Choice (D) is the result of using the

formula $P(n,r) = \dfrac{n!}{r!}$ instead of $P(n,r) = \dfrac{n!}{(n-r)!}$. Choice (E) is the value of $\dfrac{n!r!}{(n-r)!}$

Question 5: Permutations of Sets Including Like Objects

How many different strings of letters can be created by scrambling the letters of *REPEATED*?

(A) 336

(B) 2,016

(C) 6,720

(D) 13,440

(E) 40,320

REPEATED is an eight-letter word with three *E*'s. Since the letter appears more than once, some of the scrambled strings will be identical. For instance, if we scrambled the word by swapping the second and seventh letters, we'd still have the word *REPEATED*. There are six different ways to scramble the E's, all of which are identical. Let's number the E's and see, if we take the string $R E P E \underset{1}{A} T \underset{2}{E} D$, we could have

$R E P E \underset{1}{A} T \underset{2}{E} D$
$\quad \underset{1}{} \quad \underset{2}{} \qquad \underset{3}{}$

$R E P E \underset{1}{A} T \underset{3}{E} D$
$\quad \underset{1}{} \quad \underset{3}{} \qquad \underset{2}{}$

$R E P E \underset{2}{A} T \underset{1}{E} D$
$\quad \underset{2}{} \quad \underset{1}{} \qquad \underset{3}{}$

$R E P E \underset{2}{A} T \underset{3}{E} D$
$\quad \underset{2}{} \quad \underset{3}{} \qquad \underset{1}{}$

$R E P E \underset{3}{A} T \underset{1}{E} D$
$\quad \underset{3}{} \quad \underset{1}{} \qquad \underset{2}{}$

$R E P E \underset{3}{A} T \underset{2}{E} D$
$\quad \underset{3}{} \quad \underset{2}{} \qquad \underset{1}{}$

Thus, there are $\dfrac{1}{6}$ as many permutations of this word as there are permutations of a word with eight unique letters. The number of permutations of the letters of *REPEATED*, then, is $P(8,8) \div 6$, or $\dfrac{8!}{3!} = 8 \times 7 \times 6 \times 5 \times 4 = 6{,}720$. **Choice (C) is the correct answer.**

Choice (A) is the value of $P(8,3)$. Choice (B) is that value multiplied by 3!. Choice (D) gives the value of $\dfrac{8!}{3}$, rather than $\dfrac{8!}{3!}$, and (E) is simply 8!.

CHAPTER QUIZ

1. What is the value of $C(7,2)$?
   - (A) 14
   - (B) 21
   - (C) 28
   - (D) 42
   - (E) 49

2. What is the value of $P(9,5)$?
   - (A) 126
   - (B) 504
   - (C) 3,024
   - (D) 15,120
   - (E) 90,720

3. Bill has an apple, a banana, a nectarine, an orange, a pear, and a peach in a fruit basket. He will take four pieces of fruit with him. How many different selections can he make?
   - (A) 15
   - (B) 30
   - (C) 120
   - (D) 360
   - (E) 720

4. Paula must select a four-digit ID code for her computer. Each digit can be a number from 0–9, except for the first digit, which cannot be 0. How many possible ID codes can Paula choose from?
   - (A) 3,024
   - (B) 5,040
   - (C) 6,561
   - (D) 9,000
   - (E) 10,000

5. A manager of a store has three employees. Each day, one of the three employees must open up the store in the morning. The manager must make a schedule for the opening of the store each day over a five-day period. How many different possible schedules are there?
   - (A) 20
   - (B) 25
   - (C) 125
   - (D) 243
   - (E) 729

6. Eight students will each give a report in class, one at a time. In how many different orders could the students go?

   _____

7. From a group of ten committee members, one will be appointed as the chairperson, and another will be appointed as the secretary. How many possible selections are there?

   _____

8. A movie theater manager is planning to screen four different movies, from a group of nine. The manager must also decide the order in which the movies will be shown. How many possible movie schedules are there?

   (A) 126
   (B) 756
   (C) 3,024
   (D) 6,561
   (E) 15,120

9. How many different ways are there of arranging the letters of the word *MIRROR*?

   (A) 120
   (B) 240
   (C) 360
   (D) 540
   (E) 720

10. How many different ways are there of scrambling the digits of the number 9,518,191?

   (A) 210
   (B) 420
   (C) 840
   (D) 2,520
   (E) 5,040

ANSWER EXPLANATIONS

1. B

$C(7,2)$ is the number of combinations of two objects from a seven-object group. Since $C(n,r) = \dfrac{n!}{r!(n-r)!}$,

$C(7,2) = \dfrac{7!}{2!(7-2)!} = \dfrac{7!}{2!5!} = \dfrac{7 \times 6}{2} = 21$. Choice (D) is the value of

$P(7,2)$. Choice (E), which is the value of 7^2, is the number of possible two-object combinations when each object can be used twice.

2. D

$P(9,5) = \dfrac{9!}{(9-5)!} = \dfrac{9!}{4!} = 9 \times 8 \times 7 \times 6 \times 5 = 15,120$. Choice (A) is $C(9,5)$, which is the value of $P(9,5)$ divided by $5!$, or 120. Choice (C) is the

result of using the expression $\dfrac{n!}{r!}$, rather than $\dfrac{n!}{(n-r)!}$, in the permutation

formula. Choice (E) is the value of $\dfrac{9!}{4}$.

3. A

Since the question involves a group of four objects taken from a larger group of six, it is asking you to evaluate $C(6,4)$.

$C(6,4) = \dfrac{6!}{4!(4-2)!} = \dfrac{6!}{4!2!} = \dfrac{6 \times 5}{2} = 15$. Choice (B) is the value of

$P(6,2)$, and (D) is the value of $P(6,4)$.

4. D

This question allows any digit to be used more than once. So, each of the four places in the code is filled with a number from a different group, essentially. This means that factorials aren't involved here, and we won't use the standard combination or permutation formulas. Since the first digit cannot be zero, there are nine possibilities for that digit. For the other three, there are ten possibilities (0, 1, 2, 3, 4, 5, 6, 7, 8, 9). The number of possible codes, then, is $9 \times 10 \times 10 \times 10$, or $9,000$.

Choice (A) is the value of $P(9,4)$, and choice (B) is the value of $P(10,4)$. Choice (C) is the value of 9^4, and choice (E) is 10^4.

5. D

This is another combination question where each place is filled with a member from a different group, since each employee can be used any number of times. So, the number of possible combinations is the product of the number of possibilities for each spot. There are five spots, with possibilities in each, so the number is 3^5, or 243. Choice (C) is the value of 5^3; be careful when you determine which value belongs in the base and which value is the exponent.

6. 40,320

Since all eight objects of a group are being ordered here, the number of possible orders is 8!, or 40,320. Since $0! = 1$, $P(8,8) = \dfrac{8!}{(8-8)!} = \dfrac{8!}{1}$.

7. 45

This question involves making a two-object group from a ten-object group. Since each member of the two-object group has a certain position, the group is ordered. Therefore, this is a permutation question, and you need to find the value of $P(10,2)$:

$$P(10,2) = \frac{10!}{(10-2)!2!} = \frac{10!}{8!2!} = \frac{10 \times 9}{2 \times 1} = \frac{90}{2} = 45$$

8. C

This question asks for the number of permutations of four objects from a group of nine.

$P(9,4) = \dfrac{9!}{(9-4)!} = \dfrac{9!}{5!} = \dfrac{9 \times 8 \times 7 \times 6}{1} = 3,024$. Choice (B) is the value of $\dfrac{9!}{5!4}$, rather than $\dfrac{9!}{5!4!}$. Choice (A) is $C(9,4)$, rather than $P(9,4)$. Choice (D) is $P(9,5)$.

9. A

This six-letter word has one letter (R) that occurs three times. Since $3! = 6$, this set of letters has $\dfrac{1}{6}$ the number of permutations of six unique letters. $6! = 720$, so *MIRROR* has $\dfrac{720}{6}$, or 120 permutations. Choice (B) gives the value of $\dfrac{720}{3}$, rather than $\dfrac{720}{3!}$.

10. B

This seven-digit number has one digit (9) that occurs twice, and another (1) that occurs three times. Therefore, the number of unique permutations of these seven digits is $\dfrac{7!}{3!2!} = \dfrac{7 \times 6 \times 5 \times 4}{2} = 420$. Choice (C) is the number of permutations of a seven-digit number with one digit that occurs three times, but no others that occur twice. Choice (D) is the number of permutations of a seven-digit number with one digit that occurs twice, but no others that occur three times. Choice (E) is $7!$, the number of permutations of a seven-digit number with no recurring digits.

CHAPTER 12

Probability

WHAT IS PROBABILITY?

It helps to talk about probability in terms of *events*. Many events have *outcomes* that are random, at least from our point of view. The rolling of a pair of dice, for instance, is an event with a number of possible outcomes. The probability of any particular outcome is the degree of likelihood that it will happen.

CONCEPTS TO HELP YOU

Finding probabilities is essentially a matter of counting. Every probability is a ratio involving outcomes. Matters get a little more complicated when it comes to probabilities involving multiple events, or multiple ways of getting a desired outcome.

i. Probability Functions

You can think of a probability in terms of a function. A probability function doesn't necessarily pair up two numbers, however. Rather, it pairs an outcome with a number. The domain of a probability function is the set of outcomes of an event. The range is the set of probabilities of those events.

To distinguish the probability function, we'll use the letter P instead of f. The argument of a probability function is an event, which could be represented with a letter. Since the probability that a fair coin will land on tails when flipped is $\frac{1}{2}$, we could say that outcome a = coin lands on tails, and $P(a) = \frac{1}{2}$. We'll use statements like "a = coin lands on tails" or

"b = coin lands on heads" to define outcomes as arguments in probability functions. Alternatively, you could use the description of the outcome as your argument, as in $P(\text{coin lands on tails}) = \dfrac{1}{2}$.

ii. Experimental Probability

Suppose a manager at a supermarket has just inspected 50 egg cartons. He found that 8 of them have one or more broken eggs. What is the probability that the next carton he inspects will have broken eggs? Since the ratio of the number of cartons with broken eggs to the total number of cartons is 8 to 50, or $\dfrac{8}{50}$, he takes that to be the probability. In this case, he has determined an *experimental probability*. Such a probability is based on observations of prior events and their outcomes. An experimental probability represents the frequency with which a given outcome has occurred in the past.

$$\text{Experimental probability} = \frac{\text{\# of times the given outcome has occurred}}{\text{total \# of events}}$$

In the case of the egg cartons, each inspection of a carton is one event. Each of those events has an outcome: the carton has broken eggs or it doesn't. Since the first of those two outcomes occurred eight times, that's the value of the numerator in this ratio.

iii. Theoretical Probability

The *theoretical probability* of a given outcome is determined on the basis of information about different possible outcomes, as opposed to information about outcomes that have already occurred.

When the distinct possible outcomes of a given event are equally likely, the probability of a desired result can be expressed as a simple ratio:

$$\text{The probability of a desired result} = \frac{\text{\# of ways the desired result can occur}}{\text{\# of possible outcomes}}$$

When you roll a pair of dice, the numbers rolled will have a sum between 2 and 12. What is the probability of rolling a total of 2? Since there is one way of getting that outcome (rolling two 1's) out of 36 possible outcomes (we'll explain later how we got that total), the probability of rolling a total of 2 is $\frac{1}{36}$.

THE RANGE OF THE PROBABILITY FUNCTION

Since the number of ways a desired result can occur is always less than or equal to the total number of possible outcomes, the probability of a desired result cannot be greater than 1. Also, since the number of ways a desired result can occur cannot be less than 0, the probability of that result must be greater than or equal to 0. The range of the probability function $P(a)$, then, is $0 \le P(a) \le 1$.

Therefore, probabilities are expressed as fractions or decimals from 0 to 1, or as percentages from 0 to 100. An outcome with a probability of 0 or 0 percent definitely won't occur. For instance, the probability of rolling a total of 20 with a pair of standard dice has a probability of 0. An outcome with a probability of 1, or 100 percent, will definitely occur. Outcomes with higher probabilities are more likely than events with lower probabilities.

iv. The Probability That an Outcome Won't Occur

Let's take the outcome a. In any event, either a will happen or a won't happen. In any one instance, it can't be the case that a both will and won't happen, and it can't be the case that a neither will nor won't happen. Neither of those even makes sense. So, for any outcome a, $P(a$ or not $a) = 1$. Here, "not a" is the outcome in which a does *not* occur.

Since $P(a$ or not $a) = 1$, $P(a) + P($not $a) - P(a$ and not $a) = 1$. Because $P(a$ and not $a) = 0$, $P(a) + P($not $a) = 1$. Therefore, $P($not $a) = 1 - P(a)$.

This means that you can find the probability of an outcome not occurring by subtracting the probability that it will occur from 1.

From now on, we'll use $P(-a)$ instead of $P($not $a)$.

v. Simple Probability

The probability of an outcome of a single event is a *simple probability*. The probability that a fair coin will land on heads on a single flip, $\frac{1}{2}$, is a simple probability. That probability is $\frac{1}{2}$ because there is one way to get the desired outcome, out of a total of two possible outcomes.

Many questions involving simple probability will involve more than one desired outcome. Suppose you are asked for the probability of getting an odd number on a roll of a single six-sided die. There are three odd numbers you could roll: 1, 3, and 5. The probability of rolling a particular odd number is $\frac{1}{6}$; that's the probability of getting the one desired outcome out of six possible outcomes. So, the probability of rolling a 1 is $\frac{1}{6}$, the probability of rolling a 3 is $\frac{1}{6}$, and the probability of rolling a 5 is $\frac{1}{6}$. The probability of rolling any one of these three numbers is $\frac{1}{6} + \frac{1}{6} + \frac{1}{6}$, or $\frac{3}{6}$.

That is the ratio of the number of ways of getting the desired outcome (an odd number) to the total number of possible outcomes.

In many cases, if a and b are both possible outcomes of a simple event, then $P(a \text{ or } b) = P(a) + P(b)$. That is illustrated by the above example. There is one complicating factor that requires us to use a longer formula, however. We just determined that the probability of rolling an odd number is $\frac{3}{6}$, or $\frac{1}{2}$. Now, there are three prime numbers on a six-sided die: 2, 3, and 5. The probability of rolling a prime number, then, is also $\frac{3}{6}$, or $\frac{1}{2}$.

Let's say that a = rolling an odd number, and b = rolling a prime number. If $P(a \text{ or } b) = P(a) + P(b)$, then the probability of rolling either an odd number or a prime number is $\frac{1}{2} + \frac{1}{2}$, or 1. This would mean that you will definitely roll an odd or prime number, but how can that be? 4 and 6 are neither odd nor prime, and getting one of those numbers is certainly possible. Where did we go wrong?

The answer lies in the fact that 3 and 5 are *both* odd and prime. If you simply add $P(a)$ and $P(b)$ you are counting each of those outcomes twice. To avoid doing that, you need to account for the outcome a and b. Since two of the six numbers are both odd and prime, $P(a \text{ and } b) = \frac{2}{6}$. Subtracting that probability from $P(a) + P(b)$ will ensure that no outcome is counted twice.

Therefore, $P(a \text{ or } b) = P(a) + P(b) - P(a \text{ and } b) = \frac{3}{6} + \frac{3}{6} - \frac{2}{6} = \frac{4}{6}$. Since there are four numbers on a die, 1, 2, 3, and 5, that are odd and/or prime, we can be sure that the probability of $\frac{4}{6}$ reflects the number of ways of getting the desired outcome.

vi. Compound Probability

The probability of an outcome of a series of events is a *compound probability*. One of the most basic kinds of compound probability questions involves multiple coin tosses. What is the probability of a coin landing on heads both times on two flips? While the set of outcomes of a single flip is {heads, tails}, the set of outcomes of a two-flip sequence is a set of pairs: {(heads, heads), (heads, tails), (tails, heads), (tails, tails)}. The first member of each pair is the result of the first flip, and the second is the result of the second flip. That is the complete list of possible outcomes of a two-flip sequence, and one of them is the desired outcome (heads, heads). So, the probability of that outcome is $\frac{1}{4}$.

In general, given a compound event comprised of two simple events, where a is a possible outcome of one event and b is a possible outcome of the other, $P(a \text{ and } b) = P(a) \bullet P(b)$.

Let's test this with the case of three coin flips. What is the probability of getting heads on each of three coin flips? Let's list the possible outcomes in a table:

1st Flip	2nd Flip	3rd Flip
Heads	Heads	Heads
Heads	Heads	Tails
Heads	Tails	Heads
Heads	Tails	Tails
Tails	Heads	Heads
Tails	Heads	Tails
Tails	Tails	Heads
Tails	Tails	Tails

This list shows every possible combination of three flips. There are eight altogether, with only one desired outcome. There, the probability of the coin landing on heads on all three flips is $\frac{1}{8}$. The probability of getting heads on any one flip is $\frac{1}{2}$, and $\frac{1}{2} \cdot \frac{1}{2} \cdot \frac{1}{2} = \frac{1}{8}$.

vii. Dependent and Independent Events

We just explained that the compound probability $P(a$ and $b)$ equals the product of $P(a)$ and $P(b)$. Determining the value of $P(a$ and $b)$ might require more than multiplication, however. That is because whether a happens as the outcome of one event might affect how likely b is to happen. When a does affect the probability of b, the events in question are *dependent*.

Suppose a jar contains 20 red and 20 blue marbles. Someone is going to reach into the jar while blindfolded, remove one marble, and put the marble aside. She will then take out another marble. Let $a =$ removing a red marble on the first pick and $b =$ removing a red marble on the second pick. $P(a)$ and $P(b)$ are not the same!

$P(a) = \frac{20}{40}$, or $\frac{1}{2}$, since there are 40 marbles, including 20 red ones and 40 altogether, before one is picked. However, when one marble is removed and put aside, there are 39 remaining. If the first marble picked is red, then there are 19 red marbles remaining, and $P(b) = \frac{19}{39}$. If the first marble picked is blue, on the other hand, then there will still be 20 red marbles, and $P(b) = \frac{20}{39}$. Clearly, then, the value of P(b) depends on whether a occurs.

We have a way of distinguishing these probabilities. The probability function $P(b|a)$ represents the probability of b in the event of a. It is called a *conditional probability*. Here, as we already explained, $P(b \mid a) = \dfrac{19}{39}$.

Since $\dfrac{20}{39}$ is the probability that b will happen when a doesn't happen, we can say that $P(b \mid -a) = \dfrac{20}{39}$.

In this case, then, $P(a \text{ and } b) = P(a) \bullet P(b \mid a) = \dfrac{1}{2} \bullet \dfrac{19}{39} = \dfrac{19}{78}$.

Two events are *independent* when the outcome of one of them does not affect the probability of any outcome of the other. In the scenario we just used, the two simple events would have been independent if the first marble picked were immediately put back in the jar. In that case (as long as the marbles were mixed up again), the outcome of the first pick would not affect $P(b)$. $P(b \mid a)$ and $P(b \mid -a)$ would both equal $\dfrac{20}{40}$.

Another common example of independent events is a series of coin flips, which we discussed earlier. The probability of a fair coin landing on heads is $\dfrac{1}{2}$, no matter the outcome of previous flips.

STEPS YOU NEED TO REMEMBER

As you'll see, once you understand the situation in question, the solution to a probability question is likely to be straightforward. It just happens that the event in question can be very complicated. Sorting the event out, and identifying the ways in which a desired outcome might occur is not always a matter of applying a formula.

i. Identifying Events and Outcomes

Your first step in setting up a solution to a probability question should be to identify the nature of the event. Does it involve a simple or compound event? If it is a compound event, does it consist of dependent or independent simple events?

Whether you're dealing with one simple event, or several parts of a compound event, you'll want to study the question carefully in order to find the number of ways in which the outcome can occur. For instance, the outcome of getting an odd number on a roll of a die is really not just one outcome. The outcome of an odd number is really three distinct possible outcomes (1, 3, and 5). When you set up a ratio of desired outcomes to total outcomes, you need to count the number of distinct outcomes.

In many cases, there may be too many outcomes to count individually. You may have to use the methods of counting combinations and permutations we discussed in the previous chapter to find the number of possible outcomes.

ii. Using the Probability Formulas

Unless you're looking for the probability of a single distinct outcome of a simple event, you'll be looking for a probability function of the form $P(a$ or $b)$ or $P(a$ and $b)$, or something with an even longer argument, such as a or b or c, or a and b and c and d.

As explained earlier, to find the probability that a simple event will have one of two outcomes, use the formula $P(a$ or $b) = P(a) + P(b) - P(a$ and $b)$. In many cases, $P(a$ and $b)$ will equal zero, and that part of the formula won't mean any extra work.

When it comes to three or more desired outcomes, things can get complicated quickly:

$$P(a \text{ or } b \text{ or } c) = P(a) + P(b) + P(c) - P(a \text{ and } b \text{ and } -c) - P(a \text{ and } -b \text{ and } c) - P(-a \text{ and } b \text{ and } c) - P(a \text{ and } b \text{ and } c).$$

The probability of the desired outcome of a compound event is the product of simple probabilities:

$P(a \text{ and } b) = P(a) \cdot P(b)$

$P(a \text{ and } b \text{ and } c) = P(a) \cdot P(b) \cdot P(c)$

and so on.

iii. Working with Dependent Probabilities

As we've explained, the outcome of one part of a compound event can affect the probability of a desired outcome of another part. Use conditional probabilities (such as $P(b|a)$) to keep track of these effects. When finding a conditional probability $P(b|a)$, you need to determine how outcome a affects the number of ways the desired outcome can occur, as well as the total number of possible outcomes.

STEP-BY-STEP ILLUSTRATION OF THE FIVE MOST COMMON QUESTION TYPES

In order to illustrate some of the key differences among probability situations, we will go through a number of questions involving the same kind of scenario: drawing cards from a standard 52-card deck. If you're not familiar with such a deck, you only need to know a few things. There are 13 different kinds of cards in the deck: 2, 3, 4, 5, 6, 7, 8, 9, 10, Jack, Queen, King, and Ace. There are 4 cards of each kind, for a total of 52. Each of the 4 cards in a kind belongs to a different suit: Clubs, Diamonds, Hearts, or Spades. There are 13 cards of each suit, for a total of 52.

Question 1: Finding an Experimental Probability

A gumball machine contains red, purple, and orange gumballs. So far today, of the gumballs that have come out of the machine, 35 were red, 25 were purple, and 15 were orange. Based on this, which of the following is the best estimate probability that the next gumball to come out of the machine will be orange?

(A) $\frac{1}{6}$

(B) $\frac{1}{5}$

(C) $\frac{1}{4}$

(D) $\frac{1}{3}$

(E) $\frac{1}{2}$

Since this question asks you to find a probability based on observations of earlier outcomes, it is asking for an experimental probability. So far, a total of 75 gumballs have come out of the machine ($35 + 25 + 15 = 75$). 15 of those were orange. Therefore, the frequency of orange gumballs is 15 out of 75, or 1 out of 5 $\left(\frac{15}{75} = \frac{1}{5}\right)$. The experimental probability is $\frac{1}{5}$, since

$$\frac{\text{\# of times the given outcome has occurred}}{\text{total \# of events}} = \frac{15}{75} = \frac{1}{5}.$$

Choice (B) is the correct answer.

Choice (C) may be the result of using an incorrect ratio. $\frac{1}{4}$ is the ratio of the number of orange gumballs to the number of non–orange (red or purple) gumballs: $\frac{15}{35 + 25} = \frac{15}{60} = \frac{1}{4}$. Remember that probabilities and frequencies are ratios of parts to wholes, not parts to parts. Choice (D) is actually the experimental probability of a purple gumball being next. Also, one might take $\frac{1}{3}$ to be the probability by assuming that the number of gumballs in the machine is equal. That's not based on the observation of the outcomes that have already occurred, though.

Question 2: Finding the Theoretical Probability of a Simple Event

What is the probability of getting a Club card or a King from a shuffled 52-card deck on a single draw?

(A) $\dfrac{3}{13}$

(B) $\dfrac{4}{13}$

(C) $\dfrac{1}{4}$

(D) $\dfrac{17}{52}$

(E) $\dfrac{9}{26}$

This question involves a single event: the drawing of a card from a standard 52-card deck. There are 52 possible outcomes, and there are a number of ways the desired outcome can occur: drawing a 2 of Clubs, a Jack of Clubs, a King of Diamonds, and so on.

In a standard deck, there are 13 Club cards (each card in the deck belongs to one of four suits: Clubs, Diamonds, Hearts, or Spades). Therefore, the probability of getting one of those cards on a single draw is $\dfrac{13}{52}$. Since there are four Kings in a standard deck, the probability of drawing one of those is $\dfrac{4}{52}$. Now, there is one card in the deck that belongs to both groups: the King of Clubs. The probability of drawing it is $\dfrac{1}{52}$. In order to set up the solution to this, let's define these outcomes:

a = drawing a Club card
b = drawing a King
a and b = drawing the King of Clubs
or

$$P(a) = \frac{13}{52}$$

$$P(b) = \frac{4}{52}$$

$$P(a \text{ and } b) = \frac{1}{52}$$

Now we can use the probability formula for the outcome *a* and *b*:

$$P(a \text{ or } b) = P(a) + P(b) - P(a \text{ and } b) = \frac{13}{52} + \frac{4}{52} - \frac{1}{52} = \frac{17}{52} - \frac{1}{52} = \frac{16}{52} = \frac{4}{13}$$

So, choice (B) is the correct answer. Choice (A) is the value of $P(b) - P(a$ and $b)$, while (C) is just the value of $P(b)$. Choice (D) is the result of not subtracting $P(a$ and $b)$, as $P(a) + P(b) = \frac{13}{52} + \frac{4}{52} = \frac{17}{52}$. Remember that if you don't subtract $P(a$ and $b)$ when calculating $P(a$ or $b)$, you run the risk of double-counting outcomes. Finally, choice (E) uses the incorrect formula $P(a \text{ or } b) = P(a) + P(b) + P(a \text{ and } b)$.

Question 3: Finding the Probability of a Simple Event Not Occurring

What is the probability of getting neither a Diamond card nor an even-numbered card from a shuffled 52-card deck on a single draw?

(A) $\frac{19}{52}$

(B) $\frac{6}{13}$

(C) $\frac{7}{13}$

(D) $\frac{33}{52}$

(E) $\frac{47}{52}$

Let's say that a = getting a Diamond card, and b = getting an even-numbered card. The probability of getting neither a Diamond card nor an even-numbered card is the probability of getting $P(\text{not }(a \text{ or } b))$. Since $P(\text{not }(a \text{ or } b)) = 1 - P(a \text{ or } b)$, we can set up the solution to this question by finding $P(a \text{ or } b)$. That is, we can find the probability that a Diamond or even-numbered card will be drawn, and then subtract that number from 1.

Since there are 13 Diamond cards in a standard deck, $P(a) = \dfrac{13}{52}$. Now, the even-numbered cards in the deck are 2, 4, 6, 8, and 10. Since there are four cards for each of those numbers (e.g., there is a 2 of Clubs, a 2 of Diamonds, a 2 of Hearts, and a 2 of Spades), there are 20 even-numbered cards in the deck. So, $P(b) = \dfrac{20}{52}$. Note that we've already identified one card, the 2 of Diamonds, belonging to both groups. There are five even-numbered Diamond cards altogether. Therefore, $P(a \text{ and } b) = \dfrac{5}{52}$.

Now we have everything we need to find the value of $P(a \text{ or } b)$:

$$P(a \text{ or } b) = P(a) + P(b) - P(a \text{ and } b) = \frac{13}{52} + \frac{20}{52} - \frac{5}{52} = \frac{33}{52} - \frac{5}{52} = \frac{28}{52} = \frac{7}{13}.$$

Remember that we're not done yet; the question doesn't ask for the value of $P(a \text{ or } b)$. We need the value of $P(-(a \text{ or } b))$:

$$1 - P(a \text{ or } b) = 1 - \frac{7}{13} = \frac{13}{13} - \frac{7}{13} = \frac{6}{13}.$$ So, the probability that a single card drawn from a shuffled deck will be neither a Diamond card nor an even-numbered one is $\dfrac{6}{13}$.

Choice (B) is the correct answer. Choice (C) is the value of $P(a \text{ or } b)$; that is the result of not performing the last step, subtracting $P(a \text{ or } b)$ from 1. Choice (A) is the result of not subtracting $P(a \text{ and } b)$ when calculating $P(a \text{ or } b)$. Choice (D) uses that value of $P(a \text{ or } b)$. Finally, choice (E) is the value of $P(-(a \text{ and } b))$. That's that probability of drawing a card that is *not both* a Diamond and even-numbered card, as opposed to the probability of drawing a card that is *neither* a Diamond card nor an even-numbered card.

Question 4: Finding the Probability of an Independent Compound Event

One card will be drawn from a standard, shuffled 52-card deck. The card will then be put back, and the deck will be reshuffled. What is the probability that one card will be a 2 or 3 and the other will be a Queen?

(A) $\dfrac{1}{169}$

(B) $\dfrac{2}{169}$

(C) $\dfrac{4}{169}$

(D) $\dfrac{8}{663}$

(E) $\dfrac{16}{663}$

This question involves compound probability, since each of the two draws from the deck is a separate event. The desired outcome of one event is the draw of a 2 or 3, and the desired outcome of the other event is the draw of a Queen. Note, however, that the question doesn't specify the order in which these outcomes happen. So, a draw of a 2 or 3, followed by a draw of a Queen is a desired outcome of the compound event, but so is the draw of a Queen followed by the draw of a 2 or 3.

Setting up the solution to this problem involves working with combinations, which we studied in the previous chapter. That's because the number of possible outcomes here is the number of possible two-card combinations (where any card can be used twice). That number is 52 × 52, or 2,704.

This compound event is made up of a couple of simple events. The desired outcome of one of those events is a, the draw of a 2 or 3. Since there are four of each of those cards, the probability of getting one of them on a single draw is $\dfrac{8}{52}$. So, $P(a) = \dfrac{8}{52}$. If b is the other desired outcome, the draw of a Queen, then $P(b) = \dfrac{4}{52}$.

The probability of first drawing a 2 or 3 and then drawing a queen is

$$P(a \text{ then } b) = P(a) \bullet P(b) = \frac{8}{52} \bullet \frac{4}{52} = \frac{32}{2,704}.$$

That's not all, though. Remember that drawing a Queen first, and then a 2 or 3 is also a desired outcome. We just found the value of $P(b|a)$, but we need to find the probability of $P(a|b)$:

$$P(b \text{ then } a) = P(b) \bullet P(a) = \frac{4}{52} \bullet \frac{8}{52} = \frac{32}{2,704}.$$

So, the probability of drawing a 2 or 3 and then a Queen, or vice versa, is

$$P(a \text{ then } b \text{ or } b \text{ then } a) = \frac{32}{2,704} + \frac{32}{2,704} = \frac{64}{2,704} = \frac{4}{169}.$$

Choice (C) is the correct answer. You might get choice (A) by dividing $\frac{32}{2,704}$ in half instead of doubling it. Choice (B) gives the probability of just $P(b \text{ then } a)$. Choice (E) is the result of overlooking the fact that the card drawn first is put back in the deck. That would make the events dependent.

Question 5: Finding the Probability of a Dependent Compound Event

Two cards will be drawn from a standard 52-card deck, one at a time, and set aside. What is the probability that the first card drawn will be a Heart card, and the second card drawn will be a Spade card?

(A) $\frac{1}{52}$

(B) $\frac{3}{52}$

(C) $\frac{1}{17}$

(D) $\frac{1}{16}$

(E) $\frac{13}{204}$

This question involves dependent events, since the possible outcomes of the second draw are affected by the first draw. Since the first card drawn will be set aside, there will be only 51 cards.

Let a = getting a Heart card on the first draw and b = getting a Spade card on the second draw. $P(a) = \dfrac{13}{52} = \dfrac{1}{4}$, since there are 13 Heart cards in the deck. In addition to $P(a)$, we are looking for $P(b|a)$, the probability of b in the event of a. If a Heart is drawn first, then there are 13 Spades remaining, out of a total of 51 cards. So, $P(b \mid a) = \dfrac{13}{51}$.

The probability of the desired outcome, then, is
$$P(a) \bullet P(b \mid a) = \dfrac{1}{4} \bullet \dfrac{13}{51} = \dfrac{13}{204}.$$
Choice (C) is the correct answer.

Choice (B) is the result of taking $P(b|a)$ to equal $\dfrac{12}{52}$ instead of $\dfrac{13}{51}$. Choice (C) is the result of taking $P(b|a)$ to equal $\dfrac{12}{51}$ instead of $\dfrac{13}{51}$. Remember that the number of cards of one suit is not affected by the drawing of a card of another suit. Choice (D) is the result of taking the first card drawn to be returned to the deck before the second card is drawn, such that the events are independent and $P(b \mid a) = \dfrac{1}{4}$.

CHAPTER QUIZ

1. Of the visitors to a department store today, 85 went to the clothing section first, 35 went to the electronics section first, and 30 went to the furniture section first. Which of these is the best estimate of the probability that the next customer will visit the clothing section first?

   (A) $\dfrac{7}{30}$

   (B) $\dfrac{7}{23}$

   (C) $\dfrac{7}{17}$

   (D) $\dfrac{17}{30}$

   (E) $\dfrac{13}{17}$

2. Three hundred of the refrigerators assembled in a factory were inspected at random. Twelve of the inspected refrigerators were found to have a defect. Which of these is the best estimate of the probability that another refrigerator selected at random will NOT have a defect?

   (A) $\dfrac{1}{25}$

   (B) $\dfrac{1}{12}$

   (C) $\dfrac{11}{12}$

   (D) $\dfrac{24}{25}$

   (E) $\dfrac{25}{26}$

3. A class has 12 boys and 14 girls. The teacher will pick one student at random to read aloud. What is the probability that the teacher will pick a boy?

   (A) $\frac{3}{13}$

   (B) $\frac{3}{7}$

   (C) $\frac{6}{13}$

   (D) $\frac{7}{13}$

   (E) $\frac{6}{7}$

4. What is the probability of getting a total of 5 or less on a single roll of two six-sided dice?

   _____

5. What is the probability of rolling a total of 10 each time on two separate rolls of a pair of six-sided dice?

   _____

6. $P(a) = \frac{1}{3}$ and $P(b) = \frac{2}{5}$. If $P(a \text{ or } b) = \frac{2}{3}$, what is the value of $P(a \text{ and } b)$?

   (A) $\frac{1}{15}$

   (B) $\frac{2}{15}$

   (C) $\frac{1}{5}$

   (D) $\frac{7}{15}$

   (E) $\frac{3}{5}$

7. Events a and b are independent. $P(a) = \frac{1}{8}$ and $P(b) = \frac{1}{4}$. What is the value of $P(-(a \text{ and } b))$?

   (A) $\frac{9}{16}$

   (B) $\frac{5}{8}$

   (C) $\frac{21}{32}$

   (D) $\frac{15}{16}$

   (E) $\frac{31}{32}$

8. Frank will roll of pair of six-sided dice two times. What is the probability that he will get a total of 5 once, and a total of 11 once?

(A) $\dfrac{1}{162}$

(B) $\dfrac{1}{81}$

(C) $\dfrac{1}{24}$

(D) $\dfrac{1}{6}$

(E) $\dfrac{1}{3}$

9. A single card will be drawn from a 52-card deck and placed aside. What is the probability that the next card drawn will be of the same suit (Clubs, Diamonds, Hearts, or Spades) as the first?

(A) $\dfrac{1}{17}$

(B) $\dfrac{1}{16}$

(C) $\dfrac{4}{17}$

(D) $\dfrac{1}{4}$

(E) $\dfrac{33}{68}$

10. Two cards will be drawn from a standard 52-card deck, one at a time, and set aside. What is the probability that the first card drawn will be a Jack or a Club card, and the second card drawn will be a Queen or a Club card?

(A) $\dfrac{16}{221}$

(B) $\dfrac{19}{221}$

(C) $\dfrac{20}{221}$

(D) $\dfrac{81}{884}$

(E) $\dfrac{64}{663}$

ANSWER EXPLANATIONS

1. D

This question asks for an experimental probability. So far, a total of 150 people $(85 + 35 + 30)$ have come to the store. Since 85 of them visited the clothing section first, $\frac{85}{150}$, or $\frac{17}{30}$, of the visitors made that their first stop. That is the basis for an experimental probability of that value.

Choice (A) is the probability of the next visitor visiting the electronics section first. Choice (E) is the value of $\frac{30 + 35}{85}$.

2. D

This experimental probability question asks for the probability that a given event will *not* occur. If a = the next refrigerator inspected will have a defect, then $p(a) = \frac{12}{300} = \frac{1}{25}$. Therefore, $p(-a) = 1 - \frac{1}{25} = \frac{25}{25} - \frac{1}{25} = \frac{24}{25}$.

Choice (A) is the value of $P(a)$. Choice (E) is the result of taking $P(a)$ to equal $\frac{12}{300 + 12}$.

3. C

Since there are 12 boys and 14 girls, there are 26 students altogether. That is the number of possible outcomes of a random selection of one student from the entire class. Each of the 12 boys in the class represents one way to get the desired outcome in question. So, the probability of selecting one is $\frac{12}{26}$, or $\frac{6}{13}$.

Choice (B) is the result of getting a total of 28 students in the class, rather than 26. Choice (D) is the probability of selecting a girl. Choice (E) is the result of using the number of girls in the class, rather than the total number of students, in the denominator.

4. $\dfrac{10}{36}$ or $\dfrac{5}{18}$

There are 36 possible outcomes for a roll of a pair of six-sided dice. Since a desired outcome is any roll totaling 5 or less, total rolls of 2, 3, 4, or 5 are desired. There is only one way to roll a total of 2 (1's on each die). There are two ways to roll a total of 3 (a 1 and 2 or a 2 and a 1). There are three ways to roll a total of 4 (1 and 3, 2 and 2, and 3 and 1). Finally, there are four ways to roll a total of 5 (1 and 4, 2 and 3, 3 and 2, and 4 and 1). Here are the desired outcomes, listed among the possible outcomes. The first number represents that number rolled on one die, and the second represents the number rolled on the other die:

1,1	**2,1**	**3,1**	**4,1**	5,1	6,1
1,2	**2,2**	**3,2**	4,2	5,2	6,2
1,3	**2,3**	3,3	4,3	5,3	6,3
1,4	2,4	3,4	4,4	5,4	6,4
1,5	2,5	3,5	4,5	5,5	6,5
1,6	2,6	3,6	4,6	5,6	6,6

The desired outcomes, which we already listed, appear in bold. There are ten desired outcomes altogether, out of a possible 36. So, the probability of rolling a total of 5 or less is $\dfrac{10}{36}$, which can be simplified as $\dfrac{5}{18}$.

5. $\dfrac{1}{144}$

Since there are three ways to get a total of 10 on a single roll of a pair of dice (4 and 6, 5 and 5, and 6 and 4), the probability of getting such a total is $\dfrac{3}{36}$, or $\dfrac{1}{12}$. The probability that of getting that total twice in a row is $\dfrac{1}{12} \cdot \dfrac{1}{12}$, or $\dfrac{1}{144}$.

6. A

$P(a$ or $b) = P(a) + P(b) - P(a$ and $b)$. Plug in the known values and solve for the unknown:

$\dfrac{2}{3} = \dfrac{1}{3} + \dfrac{2}{5} - P(a$ and $b)$. So, $P(a$ and $b) = \dfrac{1}{3} + \dfrac{2}{5} - \dfrac{2}{3} = \dfrac{1}{15}$.

Choice (B) is the just the product of $P(a)$ and $P(b)$.

7. E

$P(-(a$ and $b)) = 1 - P(a$ and $b)$. The probability of both independent events occurring is the product of $\dfrac{1}{8}$ and $\dfrac{1}{4} \cdot \dfrac{1}{8} \cdot \dfrac{1}{4} = \dfrac{1}{32}$, so

$P(-(a$ and $b)) = 1 - \dfrac{1}{32} = \dfrac{31}{32}$.

Choice (C) is the value of $P(-(a$ or $b))$, where $P(a$ or $b) = P(a) + P(b) - P(b)$.

8. B

There are four ways to roll a total of 5 with a pair of dice (1 and 4, 2 and 3, 3 and 2, and 4 and 1). So, the probability of getting such a total is $\dfrac{4}{36}$, or $\dfrac{1}{9}$. The probability of getting a total of 11 is $\dfrac{2}{36}$, or $\dfrac{1}{18}$, because there are two ways to roll that total (5 and 6, and 6 and 5). The probability of rolling a total of 5 and then a total of 11, then, is $\dfrac{1}{9} \cdot \dfrac{1}{18}$, or $\dfrac{1}{162}$. That is also the probability of rolling a total of 11 and then a total of 5. Both outcomes are desired. The probability of getting one or the other is $\dfrac{1}{162} + \dfrac{1}{162} = \dfrac{2}{162} = \dfrac{1}{81}$.

Choice (A) is the result of counting only one of those two ways of getting a desired outcome. Choices (D) and (E) result from adding the probabilities from the first and second rolls instead of multiplying; choice (D) involves the further error of counting only one of the two ways to get the desired outcome.

9. C

There are 13 cards of each suit in a 52-card deck. When one card of a given suit is drawn from the deck, there are 12 cards of that suit remaining, out of a total of 51 cards remaining. The probability that the second card drawn will be of the same suit as the first is $\frac{12}{51}$, or $\frac{4}{17}$. Choice (A) is the result of multiplying $\frac{4}{17}$ by $\frac{1}{4}$, which is the probability of getting a card of a particular suit on the first draw. If you make such a calculation, then you must keep in mind that $\frac{1}{17}$ is the probability of just one out of four desired outcomes. There is a probability of $\frac{1}{17}$ that both cards drawn will be of a given suit. The probability that the two cards will be of the same suit (any one of the four) is $\frac{1}{17} + \frac{1}{17} + \frac{1}{17} + \frac{1}{17}$, or $\frac{4}{17}$. Choice (E) is the sum of $\frac{4}{17}$ and $\frac{1}{4}$. Remember that you need to multiply in order to find the probability of compound events.

10. D

This is a very tricky question involving dependent events. You have to keep in mind that the probability of getting the second part of the desired outcome on the second draw can vary in a number of ways, depending on what card is drawn first. Therefore, it will help to identify three ways of getting the first part of the desired outcome on the first draw, rather than just two: you can draw a card that is a Jack but not a Club (outcome *a*), a card that is a Club but not a Jack (outcome *b*), or a card that is both (outcome *c*).

$P(a) = \frac{3}{52}$

$P(b) = \frac{12}{52}$

$P(c) = \frac{1}{52}$

Let's call the outcome of getting a Queen or Club card on the second draw *d*. There are three dependent probabilities to consider.

$$P(d \mid a) = \frac{13}{51} + \frac{4}{51} - \frac{1}{51} = \frac{16}{51}$$

$\frac{13}{51}$ is the probability of drawing a Club card, given that a non-Club Jack has already been drawn; $\frac{4}{51}$ is the probability of drawing a Queen, and $\frac{1}{51}$ is the probability of drawing the Queen of Clubs.

$$P(d \mid b) = \frac{12}{51} + \frac{4}{51} - \frac{1}{51} = \frac{15}{51}$$
$$P(d \mid c) = \frac{12}{51} + \frac{4}{51} - \frac{1}{51} = \frac{15}{51}$$

Now we can find the probability of each of the three ways of getting the desired outcome: getting a non-Club Jack and then a Queen or Club, getting a non-Jack Club card and then a Queen or Club, and getting a Jack of Clubs and then a Queen or Club.

$$P(a) \bullet P(d \mid a) = \frac{3}{52} \bullet \frac{16}{51} = \frac{48}{2,652}$$

$$P(b) \bullet P(d \mid b) = \frac{12}{52} \bullet \frac{15}{51} = \frac{180}{2,652}$$

$$P(c) \bullet P(d \mid c) = \frac{1}{52} \bullet \frac{15}{51} = \frac{15}{2,652}$$

The probability of getting one of these three desired outcomes is

$$\frac{48}{2,652} + \frac{180}{2,652} + \frac{15}{2,652} = \frac{243}{2,652} = \frac{81}{884}$$

Choice (B) is the sum of $P(a) \bullet P(d|a)$ and $P(b) \bullet P(d|b)$. Choice (C) is the result of taking $P(d)$ to be $\frac{15}{51}$ in any event, while choice (E) is the result of taking $P(d)$ to be $\frac{16}{51}$ in any event.

NOTES

NOTES

NOTES

NOTES

NOTES

NOTES

NOTES

NOTES

NOTES

NOTES